# The Social Consequences of Climate Change

# The Social Consequences of Climate Change: Debates in Research and Policy


EDITED BY

**ŞUAY NİLHAN AÇIKALIN**

*Ankara Hacı Bayram Veli University, Turkey*

AND

**ŞEFIKA ŞULE ERÇETIN**

*Hacettepe University, Turkey*



emerald
PUBLISHING

United Kingdom – North America – Japan – India – Malaysia – China

Emerald Publishing Limited
Emerald Publishing, Floor 5, Northspring, 21-23 Wellington Street, Leeds LS1 4DL

First edition 2025



**British Library Cataloguing in Publication Data**
A catalogue record for this book is available from the British Library

ISBN: 978-1-83797-678-2 (Print)
ISBN: 978-1-83797-677-5 (Online)
ISBN: 978-1-83797-679-9 (Epub)

Printed and bound by CPI Group (UK) Ltd, Croydon, CR0 4YY

*To the countless individuals and communities already experiencing the social consequences of climate change. Your resilience and strength inspire us…*

*This book is written for you and for a future where these struggles are lessened, where communities thrive, and where justice prevails.*

*Şuay Nilhan Açıkalın*
*Şefika Şule Erçetin*

# Contents

# About the Editors

**Assoc Prof Dr Şuay Nilhan Açıkalın**, Ankara Hacı Bayram Veli University, Turkey. She is a faculty member in the Department of International Relations at Ankara Hacı Bayram Veli University and a Visiting Fellow in Mathias Corvinius Collegium, Hungary. Dr Açıkalın graduated from the Department of International Relations at Bilkent University in 2012. She received her Master's degree with the thesis "Leadership in Chaos: Angela Merkel and Euro Crisis" at the Middle East Technical University in 2015. She obtained her PhD degree in 2020 from the same university with her thesis "Political Leadership and Foreign Politics: Recep Tayyip Erdoğan and Angela Merkel". She has been a guest researcher at Lancaster University in the United Kingdom and Humboldt Universitat zu Berlin in Germany within Jean Monnet Scholarship Program. Dr Açıkalın works on chaos and complexity theory, leadership, Turkish foreign policy, Turkish-EU relations, women and girls in migration studies, climate change as well as gastrodiplomacy and fashiondiplomacy in theory and practice. She is the author of the book *Leader to Leader Diplomacy in Turkish-German Relations* as well as various international books. Also, Açıkalın is the co-editor of a forthcoming book called *The Social Consequences of Climate Change: Debates in Research and Policy* by Emerald.

She is a member of the Advisory Board of Turkish Healthy Cities Union. In addition to this, she is a founding member of International Science Association, and she has been working as an expert in Diplomacy Foundation. Dr Açıkalın is a member of Bali Democracy Forum and World Indonesianist Congress. She is constantly been giving interviews to national and international media outlets such as TRTWORLD, Al Jazeera, Phoenix TV, etc.

**Prof Dr Şefika Şule Erçetin**, Dean of Faculty of Education, Hacettepe University, Türkiye. She is currently the Dean of Education Faculty of Hacettepe University, Ankara, Türkiye. In 1991, she completed her PhD from Division of Educational Administration, Supervision, Planning, and Economics in Hacettepe University. Since 2003, she has been working as a Professor. She worked as a Visiting Scholar in Texas A&M, USA, and Lancaster University, UK. She has worked in many national and international projects as a member of the project team and the project coordinator on education of refugees, women, and vocational training. She was the Founder of International Symposium of Chaos, Complexity and Leadership. She is one of the rare academicians in Turkey who interests are in "chaos, complexity, quantum leadership" since the 90s. Also, she is the author and editor of books on complexity theory from Springer, World Scientific, Apple Academic Press related with comparative education policies, research methods, organizational intelligence and stupidity and wisdom, leadership, migration, fuzzy logic, children at risk, women and peace studies, social media, plasma leadership, quantum leadership, women leadership model. Also, she is the President, Board Member, and Founder member of the International Science Association in Türkiye (ISCASS). She is the Editor-in-Chief of different national and international journals which focus on leadership and education.

# About the Contributors

**Dr Tjandra Yoga Aditama** is currently the Director of the Graduate School at YARSI University in Jakarta, Indonesia, where she is also a Professor at the Faculty of Medicine, University of Indonesia. He makes significant contributions as a member of the Expert Team of the Advisory Board to the President of the Republic of Indonesia and as a key member of the IAVG-COVAX initiative. His extensive career includes notable leadership roles, such as his tenure as the Act Director of Communicable Diseases at the World Health Organization's South-East Asia Regional Office and as the Director General of Communicable Diseases and the Head of National Institute of Health Research and Development within the Ministry of Health, Indonesia. His influence extends beyond national borders, as demonstrated by his advisory positions with international organizations like the Asia Pacific Cities Alliances for Health and Development (APCAT) and his contributions to the Australia Indonesia Health Security Partnership (AIHSP). With a remarkable publication record comprising over 200 scientific articles in medical journals, 254 popular scientific health articles, and 38 medical and health books, Prof Aditama's scholarly contributions have significantly enriched the field of public health.

**Mustafa Başkara** received his LLB from Ankara University in 2013 and his LLM in International Shipping Law from Queen Mary University of London in 2016. He continues his PhD in Maritime Law at Ankara University. He was an intern at International Maritime Organization in 2016. He worked as a Research Assistant at the Faculty of Law, Social Sciences University of Ankara from 2017 to 2021. Between 2020 and 2022, he served as an Advisor to the Minister of Justice of the Republic of Türkiye. Mr Baskara has been the CEO of Ankara University National Center for the Sea and Maritime Law since January 28, 2021, and the Director of the Mediterranean Climate Change Research Center (AİDAM) since 2023, which will gain the status of Regional Activity Center on Climate Change under the UNEP/MAP Barcelona Convention.

**Assoc Prof Dr Kaan Batı** received his Bachelor's degree in Science Education from Hacettepe University in 2005, his Master's degree in Science Education in 2010, and his PhD degree in Science Education in 2014 from Hacettepe University. He prepared his Master's thesis on scientific process skills and problem-solving and his doctoral thesis on modeling, nature of science and critical thinking. In the 2018–2019 academic year, he completed his postdoctoral studies at the Ohio State University (USA) within the scope of TÜBİTAK 2219A project.

**Brígida Brito**, Associate Professor, Deputy Director and Head of Quality at the Department of International Relations, Universidade Autónoma de Lisboa (Portugal), Deputy Director of OBSERVARE, and Editor-in-Chief of JANUS.NET, e-journal of international relations. She is an integrated Researcher at the Portuguese Institute of International Relations of the New University of Lisbon (IPRI-NOVA). She is a Sociologist and has a PhD in African Studies and a Master's degree in the same scientific field. She completed a postdoctorate in socio-environmental studies in Africa. She participates in and organizes scientific meetings and has numerous articles and chapters in reputable publications.

**Büşra Deniz** completed her undergraduate studies at Ankara University Faculty of Law in 2020. Since 2021, she has been pursuing her Master's degree in the program "The Sea and Maritime Law," conducted in English at Ankara University Institute of Social Sciences. In addition to her academic pursuits, Büşra Deniz serves as a Researcher at DEHUKAM, contributing to various projects and studies. Currently, she is involved as a Marine Environmental Law Expert in the "Project for Capacity Development in Marine Environment Law within the Framework of the Barcelona Convention and Other International Documents" and the "The Project for Exploring Potential Natural Sites with the Aim of Increasing the Protected Areas in Türkiye" in collaboration with the Ministry of Environment, Urbanization, and Climate Change. Büşra Deniz is proficient in advanced-level English.

**Dr Kübra Ecer**, who began her professional career as a Social Security Auditor in 2009 and served in this role for 10 years, was appointed as a Communications Specialist at Presidency of the Republic of Türkiye Directorate of Communications in 2019, where she continues to hold this position. Holding Master's degrees in European Union Studies and Neuromarketing, as well as a PhD in Economics, Ecer has published numerous academic studies in the fields of European Union affairs, climate crisis, sustainability, and neuroscience. Recently, she has been conducting experimental studies on the use of neuroscience techniques in climate change and sustainability, continuing her research and projects on methods to enhance sustainability communication. Additionally, she is a member of the working group for the COST project titled "Social Sciences and Humanities for Transformation and Climate Resilience-CA21166."

**Metin Erol** graduated from Gediz University, Department of Political Science and Public Administration, which he entered with a full scholarship in 2010, with first place and summa cum laude in 2015. In the same year, he completed his minor program at Gediz University, Department of International Relations. He completed his Master's degree in the Public Policy Program of Marmara University, Department of Political Science and Public Administration, with his thesis titled Struggle for Cultural Hegemony in Turkish Politics: Comparative Analysis of the Single Party and AK Party Periods. Erol continues his doctoral studies at Istanbul University, Faculty of Political Sciences, Department of Political Science and International Relations. Erol worked as a Research Assistant at the Politics,

Economics, and Society Research (SETA) Foundation between April 2016 and June 2021. From September 2021 to July 2022, he served as the Deputy Chief Correspondent for Anadolu Agency Analysis News. As of July 27, 2022, Erol serves as the Istanbul Regional Manager of The Republic of Türkiye Directorate of Communications.

**Assoc Prof Oğuz Güner** is a faculty member at the Ankara Yıldırım Beyazıt University Department of International Relations. Having completed his undergraduate degree at Yıldız Teknik University, he earned his Master's and PhD from İstanbul University Institute of Social Sciences Department of European Union Studies. He was a Visiting Fellow at Aalborg University Faculty of Social Sciences (Denmark) in 2018. He then went on to serve as a Postdoctoral Researcher in 2019 at Strasbourg University (France) Societies, Actors, Government in Europe Research Center (Sociétés, Acteurs, Gouvernement en Europe) as part of the Jean Monnet Scholarship Program. Güner has papers on the European Union, the Balkans, and EU – Türkiye relations. He lectures to undergraduate and graduate students at various universities. Assoc Prof Güner currently serves as the Head of the Department of Public Diplomacy at the Republic of Türkiye, Directorate of Communications. He is also a member of the UNESCO Türkiye National Commission Communication Specialization Committee.

**Klaus Jurgens** holds a postgraduate degree from London School of Economics and Political Science in Government Studies; member of German Association of Press Journalists; Director of Economyfirst Limited London, a media representation services company; member of More Speakers – The Speaker Agency. For this book contribution extensive research on location about Smart Cities as well as sustainable neighborhoods was carried out with a special focus on Sweden; besides, focus on Türkiye's 360 degrees foreign policies; Branding of the Nation State; The future of NATO. Frequent Guest Speaker on international television, Op Ed and Analysis Writer for various media in Türkiye and for TRT Deutsch Berlin; previously lectured about EU Law for Business Administrators in Ankara. Recent speaker engagements: International Investors Forum about Azerbaijan and sustainable energy in Brussels 2023; Azerbaijan Shusha Global Media Forum 2023; Directorate of Communications of the Presidency of the Republic of Türkiye, Stratcom 2023, Istanbul; Azerbaijan 2nd Shusha Global Media Forum 2024.

**Burak Kaplan**, in 2014, started his undergraduate education at Eskişehir Osmangazi University, Department of Political Science and Public Administration. In 2017, he graduated with the first rank in the department and continued his postgraduate education at Marmara University Public Policy Master's Program. He defended his thesis titled "The importance of innovation-based security strategies in achieving urban sustainability in the context of industry 4.0: The case of Gölcük" in 2019. Kaplan continues his doctoral studies at Marmara University, Department of Local Governments. Kaplan worked as a Lecturer at Istanbul Beykent University between 2019 and 2021 and as a Research Assistant between 2021 and 2023. Since September 4, 2023, he has been working as an Assistant

Specialist at the Istanbul Regional Directorate of the Presidency of Communications of the Republic of Türkiye.

**Serdar Karagöz**, AA President and Chief Executive Officer, born on May 24, 1981, in Sakarya, he graduated from the Department of Political Science at Istanbul Bilgi University, where he later completed his Master's degree in Human Rights Law. Karagöz earned his second Master's degree in Media Studies at New School University in the United States. His career began as a Researcher at the ADAM Strategic Research Center in 2004. In 2008, he joined the Turkuvaz Media Group as an Executive candidate and worked in all departments of the group. He joined the New York Times Syndication Department in 2010, followed by his position as UN Diplomacy Correspondent for A Haber Television in 2011–2013. Upon his return to Türkiye, he worked as the Foreign News Manager at Sabah newspaper. During this period, he also began preparations to establish the English-language Daily Sabah. In February 2014, he became the founding Editor-in-Chief of Daily Sabah. Authoring numerous columns in both English and Turkish, he launched the website usabah.com, where he monitored the pulse of events in the United States. In 2018, he was appointed Editor-in-Chief of TRT International News Channels. During his time in this position, he added Arabic, German, Russian, and Balkan languages to TRT's list of broadcasting languages and contributed significantly to its success in the global media arena. In 2020, he continued his duty as TRT Deputy General Director. Since April 7, 2021, Karagöz has been serving as the President and Chief Executive Officer of Anadolu Ajansı, Türkiye's global news agency.

**Abhinav Mehrotra**, Assistant Professor, Jindal Global Law School, O. P. Jindal Global University, Sonipat, India. Abhinav Mehrotra is an Assistant Professor at Jindal Global Law School. He teaches courses on Human Rights and Public International Law to law students. His most recent publications focusing on justifying the Right to Internet Access as a Human Right have been in the prestigious Comparative Law Review, and another article titled "Need for a National Legislation on Refugees in India at 75" was published in India Quarterly as well as the article titled "Crawford, Twail, and Sovereign Equality of States: Similarity and Differences" has been published in the Australian Yearbook of International Law volume 40 (2022), and a book chapter that focuses on the utility of the European Mechanisms on Human Rights in South Asia has been published by Springer. He recently published a coauthored journal article in the *Manchester Journal of Transnational Islamic Law and Practice*.

**Prof Dr Vesselin Popovski** is the Founding Executive Director of the Center for the Study of the UN, Vice Dean (Institution Building) of the Jindal Global Law School, Senior Advisor of the International Institute for Higher Education and Research, and Advisory Council Member of the Office for Inter-Disciplinary Studies at O.P. Jindal Global University. In 2004–2014, he was a Senior Academic Officer, Head of "Peace and Governance" Program at UN University, Tokyo. Prior to that, Co-Director of the EU project "Legal Protection of Individual Rights in Russia" and Assistant Professor, Exeter University, United Kingdom. He has a PhD from King's

College London; MSc. from London School of Economics; and BA/MA from Moscow Institute of International Affairs. He has written or edited over 25 books and numerous articles in academic journals.

**Büşra Sarı** is an energetic and future-oriented Teacher and Researcher residing in Sakarya. She graduated from Selçuk University, Department of Child Development and Education with the first rank. After her graduation, she started her teaching career, completed her Master's degree at Sakarya University, Department of Educational Administration in order to expand her knowledge and experience, and then started her doctoral studies in the same department at Hacettepe University. In addition to her work in the Department of Education, she conducts research on important topics such as leadership, educational management, and climate change. However, she is not limited to the academic world; her personal interests include photography, cycling, and visual design. Buşra SARI strives to maintain a balance between maintaining the seriousness of her profession while not neglecting to explore the joys of life.

**Assoc Prof Dr Duygu Sönmez** has been serving as a faculty member at Hacettepe University, Department of Science Education, since 2007. She holds a Bachelor's of Science in Biology from Ege University. She completed her Master's and Doctor of Philosophy degrees in Science/Biology Education at the Ohio State University. She has publications both in national and international journals and books such as the *International Journal of Science Education*, *The American Biology Teacher*, and *International Research in Geographical and Environmental Education*. Her research interests focus on biology education, science education, climate change education, teacher professional development, STEM education, and educational technology. Dr Duygu Sönmez participated in projects both in the United States and Türkiye as a Researcher and Project Manager. She continues to serve at Hacettepe University in an administrative capacity as well.

**Derya Ulutürk** completed her undergraduate studies at Başkent University Faculty of Law in 2019. Since 2021, she has been pursuing her Master's degree in the "Sea and Maritime Law" program at Ankara University Institute of Social Sciences, which is conducted in English. Working as a Researcher within DEHUKAM, Ulutürk is actively engaged in various activities related to education, conferences, and projects within the institution. Currently, she serves as a Marine Environmental Law Expert in the "Project for Capacity Development in Marine Environment Law within the Framework of the Barcelona Convention and Other International Documents" and the "The Project for Exploring Potential Natural Sites with the Aim of Increasing the Protected Areas in Türkiye" conducted in collaboration with the Ministry of Environment, Urbanization and Climate Change. Derya Ulutürk is proficient in advanced English and lower intermediate German.

**Amit Upadhyay**, Associate Professor, Jindal Global Law School, O. P. Jindal Global University, Sonipat, India. Amit Upadhyay is currently working as an Associate Professor at Jindal Global Law School. He graduated with a BA, LLB (Hons) degree from Amity Law School, New Delhi. He was awarded a prestigious DAAD scholarship to pursue a Master's degree in International Business, Competition, and Regulatory Law from Freie Universität, Berlin, Germany. He recently published a coauthored journal article in the *Manchester Journal of Transnational Islamic Law and Practice*. His academic and research areas are competition law, international law, human rights law, and constitutional law.

**Ellen Wasylina**, a multicultural leader who has spent her entire career in France, is a renowned Geopolitical and Geoeconomic Expert, with a true vision and demonstrated leadership in Sustainable Development, be it at the city level (Director of Environment and Sustainable Development), leading a think tank (International Geostrategic Maritime Observatory), being a Senior Expert for a EuropeAid consortium (Tlemcen, Algeria), or teaching in Parisian business schools. She is also a respected author, speaker, commentator, and most recently a podcast producer (Mediterranean Sustainability Partners), reaching audiences in 76 countries in English and French. Her latest endeavor is to train a new generation of leaders in sustainable leadership and diplomacy (Trocadéro Forum Institute). Her life's work and expertise has been largely influenced, inspired by her forefathers in Eastern Europe, from the former Soviet Union sphere via the Caspian, Black, and Mediterranean Seas. With a love for travel, she is passionate about music, Renaissance, and Impressionist arts, fashion, cuisine, and wine. She recently started to write a series of books, entitled "Vignes et Mets: Sustainable and Seasonal Cuisine."

# Preface

When we are approaching the end of the first quarter of the 21st century, humanity faced numerous challenges in global scale. Unfortunately, the world is under the shadow of the Russian–Ukrainian war and Palestine–Israeli conflict trying to adjust itself to untraditional challenges such as digital threats and climate change. I can imagine the feeling when you first read the word of climate change, still suspicious or somehow boring. Even some of the reader audience raised the question: Is climate change real?

This book focuses on the global, social, cultural, and economic problems of climate change and discusses the solutions and steps to be taken to address global climate change. The book contains 13 chapters dealing with different contexts of climate change. In the first chapter, Popovski examines the various impacts of climate change, including its effects on vulnerable communities, public health, migration patterns, economic disparities, and political stability. The chapter argues that combating climate change and addressing its social impacts should be a distinct purpose of the United Nations Organization, in addition to maintaining international peace and security, promoting economic and social development, and respecting and protecting human rights.

Upadhyay and Mehrotra analyze whether a human rights approach can be used to address climate change issues in India. They discuss the debate on the compatibility of aims in addressing climate change and realizing human rights, focusing on climate change issues in Uttarakhand, India. The text discusses the lack of effective adaptation policies and how this impacts the social and economic rights of those living in affected regions. The authors conclude that any strategy aimed at addressing climate change, whether through adaptation or mitigation, must consider the impact on individuals and communities, highlighting the significance of the human rights framework. Erçetin and Sarı present a metaphor to illustrate the dynamic interactions involved in climate change education. They focus on adapting Newton's cradle to the field of climate change education. Açıkalın provides a literature examination on how public policy can be more effective in combating climate change by applying complexity theory in two dimensions: reforming national policies and renewing cosmopolitanism on a local to global scale. By exposing the fragility of traditional models of governance of climate change, Brito set out on a bibliographic survey to enable a rediscovery of practices that promote responsible and socio-environmental justice at the international level. He focusses on conceptual and documentary definitions and presented cases illustrating the socio-environmental impacts of extreme weather events that can lead to or exacerbate forced displacement for environmental reasons. Recognizing that education is one of the most powerful tools we have in combating climate change, Sönmez recommended the use of systems thinking in climate change education and education for sustainable development and proposed a framework in this context. Parallel with Sönmez, Batı discusses climate change, its potential problems, the effects of climate change in Turkey, policies implemented to address it, public awareness of climate change, and teacher education in the context of climate change in Turkey.

Jurgens discusses the "build, shop, work differently" approach and its potential to implement self-contained and sustainable neighborhoods. This can reduce unnecessary daily individual traffic and avoidable pollution. The author analyzes the use of technology and emphasizes the urgency of individual citizen responsibility. Erol and Kaplan examine the compliance of Istanbul Metropolitan Municipality's (IMM) strategic plans for the periods 2015–2019 and 2020–2024 with the 17 Global Development Goals of the United Nations. Their aim is to compare the compatibility of both plans, which were prepared with different political understandings, with the 17 Global Development Goals. The study aims to codify the aims and objectives of IMM's strategic plans in different periods and assess their compliance with the 17 Global Development Goals. Ecer and Güner analyze the measures taken by the EU to address environmental concerns and the impact of climate change within the framework of the EGD. They aimed to evaluate the EU's actions toward achieving a more sustainable environment, with a significant reduction in the negative effects of climate change and harmful gas emissions. Finally, Deniz, Ulutürk, and Başkara examine Turkey's efforts to address climate change both domestically and internationally. They analyze Turkey's climate change policies and legislative developments, with a focus on the government's agenda, legal and institutional frameworks. They also provide a comprehensive overview of Türkiye's political progress on climate change at both the national and international levels.

Editors
Şuay Nilhan Açıkalın and Şefika Şule Erçetin

Part 1

# Old Topic With New Narratives

# Chapter 1

# Air Pollution

*Tjandra Yoga Aditama*

Universitas YARSI, Indonesia

## Abstract

This article examines the interconnections between climate change, air pollution, and their impacts on human health and the environment, with a focus on the Indonesian context. As one of the most biodiverse countries in the world, Indonesia faces significant challenges arising from climate change and air pollution, exacerbated by factors such as rapid urbanization, industrialization, and deforestation. The article discusses the various sources of air pollution in Indonesia, including vehicular emissions, industrial activities, biomass burning, and agricultural practices, and their contributions to greenhouse gas emissions. Furthermore, it explores the health implications of exposure to air pollutants, such as respiratory diseases, cardiovascular problems, and adverse birth outcomes, as well as the environmental consequences, including deforestation, loss of biodiversity, and disruptions to ecosystems. Finally, the article highlights the importance of policy interventions, technological innovations, and public awareness campaigns in mitigating air pollution and addressing the complex challenges posed by climate change in Indonesia.

*Keywords*: Climate change; air pollution; human health issues based on air pollution; environment; industrial activities

Air pollution is a lethal toll, and the evidence is indisputable. Air pollution is one of the greatest environmental risks to health, and it is a contamination of the indoor or outdoor environment by any chemical, physical, or biological agent that modifies the natural characteristics of the atmosphere. The repercussions of exposure to air pollution span a wide spectrum of health effects. These impacts can be delineated into immediate and enduring consequences. Short-term ramifications, which are temporary in nature, comprise ailments such as pneumonia or bronchitis. They also encompass discomfort, including irritation of the nasal

**The Social Consequences of Climate Change, 3–10**

**Published under exclusive licence by Emerald Publishing Limited**
**doi:10.1108/978-1-83797-677-520241001**

passages, throat, eyes, or skin. Additionally, air pollution can trigger symptoms such as headaches, dizziness, and queasiness. Malodorous emissions emanating from industrial facilities, waste disposal sites, or sewage systems are also classified as air pollution, albeit of a lesser magnitude. Long-term effects of air pollution can persist for extended periods, potentially even resulting in mortality. Chronic health conditions associated with air pollution encompass cardiovascular diseases, lung cancer, and respiratory disorders like emphysema. Furthermore, air pollution has the capacity to inflict enduring damage on the nervous system, brain, kidneys, liver, and other vital organs. Certain researchers posit that air pollutants may contribute to the occurrence of birth defects. Responses to various forms of air pollution vary among individuals, with young children and older adults, often possessing weaker immune systems, being particularly susceptible. Preexisting conditions such as asthma, cardiac ailments, and pulmonary diseases can be exacerbated by exposure to air pollution. Duration of exposure and the composition of pollutants are additional factors influencing the severity of health impacts (National Geographic, 2023).

Air pollution and climate change are intertwined phenomena, sharing common origins and interconnected impacts. The emissions of major pollutants not only degrade air quality but also play a significant role in altering the Earth's climate. Moreover, the sources of these pollutants often coincide with those of greenhouse gases, exacerbating the problem. WHO (2023a) stated that many of the drivers of air pollution (i.e. combustion of fossil fuels) are also sources of greenhouse gas emissions. Policies to reduce air pollution, therefore, offer a win-win strategy for both climate and health, lowering the burden of disease attributable to air pollution, as well as contributing to the near- and long-term mitigation of climate change. Air pollution is part of a web of global challenges, converging on health. The combustion of fossil fuels in transport, energy and industry sectors, the lack of clean, sustainable energy for households and health-care facilities, emissions from agriculture or mismanagement of waste and unsustainable land-use planning result in high levels of air pollution, intensifying the climate emergency and consequently ill health. It is a social, environmental and health crisis. The concept of global warming denotes a climatic shift characterized by elevated temperatures in both the atmosphere and oceans, influenced by both natural forces and human-generated air pollutants. This phenomenon is spurred by the amplification of greenhouse gas levels in the atmosphere, which, in turn, act as thermal insulators, ensnaring heat energy within the Earth's atmospheric envelope (National Geographic, 2023).

## Ambient Air Pollution

Global data show that more than 90% of the population breathe air that exceeds WHO guideline limits and contains high levels of pollutants, with low- and middle-income countries suffering from the highest exposures. The issue of outdoor air pollution stands as a significant environmental and public health concern affecting populations across the globe, regardless of economic status. Both urban and rural areas are impacted by ambient air pollution, which was estimated to result in approximately 4.2 million premature deaths worldwide in 2019 alone. This mortality rate is primarily linked to exposure to fine particulate matter (PM), contributing to the development of cardiovascular diseases, respiratory ailments, and various cancers. According to data from the World Health Organization (WHO), in 2019, around 37% of premature deaths attributable to outdoor air pollution were attributed to ischemic heart disease and stroke, while chronic obstructive pulmonary disease (COPD) and acute lower respiratory infections accounted for 18% and 23% of deaths, respectively. Additionally, respiratory tract cancers were responsible for 11% of these fatalities (WHO, 2022).

Residents of low- and middle-income nations bear a disproportionate share of the consequences of outdoor air pollution, with 89% of the estimated 4.2 million premature deaths occurring in these regions. The WHO South-East Asia and Western Pacific Regions face the heaviest burden. Recent assessments underscore the substantial impact of air pollution on cardiovascular health, highlighting its role in cardiovascular diseases and related fatalities. By mitigating air pollution levels, countries have the potential to alleviate the burden of various ailments, including stroke, heart disease, lung cancer, chronic respiratory diseases, and acute respiratory infections such as asthma. Enhancing air quality not only fosters improvements in public health but also facilitates socioeconomic development and promotes environmental well-being.

### *Pollutants*

Air pollution encompasses a multitude of chemicals or particles suspended in the air, posing threats to the health of humans, animals, and plants, as well as causing harm to structures. Pollutants in the atmosphere manifest in various forms, including gases, solid particles, and liquid droplets. Pollution infiltrates the Earth's atmosphere through diverse pathways, with a significant portion stemming from human activities. Emissions from factories, vehicles, aircraft, and aerosol cans represent major anthropogenic sources of air pollution. Additionally, secondhand cigarette smoke is recognized as a significant contributor to air pollution. Conversely, certain types of air pollution, such as smoke from wildfires or ash from volcanic eruptions, occur naturally and are categorized as natural sources of pollution.

Air pollution tends to be particularly prevalent in densely populated urban areas, where emissions from numerous sources accumulate. In some instances, geographical features such as mountains or tall buildings can impede the dispersion of pollutants, resulting in the formation of a visible haze known as smog. Smog, characterized by a murky appearance, is a common occurrence in such locations. The term "smog" is derived from the combination of "smoke" and "fog," reflecting its composition and appearance (National Geographic, 2023).

There are several important air pollutants that impacted human health and environment. PM serves as a widely used proxy indicator for assessing air pollution levels. Extensive research has provided compelling evidence regarding the adverse health effects linked to exposure to this pollutant. PM comprises various components, including sulfates, nitrates, ammonia, sodium chloride, black carbon, mineral dust, and water. These diverse constituents contribute to the overall composition of PM and play crucial roles in influencing air quality and its associated health impacts.

Beside particulate matter, there are also several gasses. Carbon monoxide (CO) is a toxic gas that is both colorless and odorless, making it imperceptible to human senses. It is generated through the incomplete combustion of carbon-based fuels like wood, gasoline, charcoal, natural gas, and kerosene. Ozone ($O_3$) at ground level, distinct from the protective ozone layer in the upper atmosphere, serves as a key component of photochemical smog. Ground-level ozone forms as a result of chemical reactions involving precursor gases under the influence of sunlight. Nitrogen dioxide ($NO_2$) is a gas commonly emitted during the combustion of fuels in transportation and industrial activities. Last but not least, Sulfur dioxide ($SO_2$) is a colorless gas with a sharp odor. It is produced from the burning of fossil fuels (coal and oil) and the smelting of mineral ores that contain sulfur (WHO, 2022).

The WHO Global air quality guidelines (AQG) serve as a comprehensive framework offering global recommendations on thresholds and limits for critical air pollutants associated with health risks. These guidelines are developed with rigorous methodology and transparency, relying on evidence-based decision-making processes. Alongside guideline values, the WHO Global AQG include interim targets aimed at facilitating a gradual transition from elevated to lower pollutant concentrations. Moreover, the guidelines provide qualitative guidance on best practices for managing specific types of PM, such as black carbon/elemental carbon, ultrafine particles, and particles originating from sand and dust storms. For these PM categories, where quantitative evidence is insufficient for establishing specific AQG levels, qualitative statements are offered to guide effective management strategies (WHO, 2022).

## Household Air Pollution

In addition to ambient air pollution, household air pollution poses significant health risks, contributing to an estimated 3.2 million deaths annually as of 2020, with over 237,000 fatalities occurring among children under the age of 5. Approximately 2.3 billion individuals worldwide, roughly one-third of the global population, rely on open fires or inefficient stoves fueled by kerosene, biomass (including wood, animal dung, and crop waste), and coal for cooking purposes, thereby generating harmful household air pollution. Exposure to household air pollution is associated with the development of noncommunicable diseases such as stroke, ischemic heart disease, COPD, and lung cancer. Women and children, who often undertake household tasks such as cooking and gathering firewood, bear the brunt of the health burden stemming from the use of polluting fuels and technologies within households (WHO, 2023b).

Heating a household using substances like kerosene, wood, and coal can result in indoor air pollution. The emission of ash and smoke from these sources not only makes breathing challenging but also leaves residues that adhere to surfaces such as walls, food, and clothing. Moreover, naturally occurring radon gas, a carcinogenic substance, can accumulate within homes. Radon emanates from the Earth's surface and can pose significant health risks. However, affordable systems, when installed by professionals, can effectively mitigate radon levels. Certain construction materials, including insulation, can also pose health hazards to occupants. Additionally, inadequate ventilation within homes and individual rooms can promote the proliferation of toxic mold. Even a solitary colony of mold thriving in damp and cool areas of a house, such as between walls, can release spores into the air, leading to their dispersion throughout the entire household. Inhalation of these spores can result in adverse health effects (National Geographic, 2023).

Among the 3.2 million deaths attributed to household air pollution exposure, 32% are attributed to ischemic heart disease. Furthermore, data indicate that 12% of all deaths caused by ischemic heart disease, totaling over a million premature deaths annually, can be directly linked to exposure to household air pollution. After heart diseases, 23% of deaths are from stroke. Approximately 12% of all deaths due to stroke can be attributed to the daily exposure to household air pollution arising from using solid fuels and kerosene at home. In the context of lung diseases, among deaths associated with household air pollution, 21% are attributed to lower respiratory infections, 19% to COPD, and 6% to lung cancer. Exposure to household air pollution significantly increases the risk of childhood lower respiratory tract infections (LRI), nearly doubling the likelihood. Moreover, household air pollution accounts for 44% of all pneumonia deaths in children under the age of 5. Additionally, household air pollution poses a risk for acute lower respiratory infections in adults and contributes to 22% of all adult deaths resulting from pneumonia. Data also showed that 23% of all deaths from COPD in adults in low- and middle-income countries are due to exposure to household air pollution, and approximately 11% of lung cancer deaths in adults are attributable to exposure to carcinogens from household air pollution caused by using kerosene or solid fuels like wood, charcoal or coal for household energy needs (WHO, 2023b).

## Effects to the Environment

Similar to humans, animals, and plants, entire ecosystems can suffer adverse effects from air pollution. Haze, akin to smog, constitutes a visible form of air pollution that obscures shapes and colors, even dampening sounds. Particles of air pollution eventually settle back onto the Earth's surface, directly contaminating bodies of water and soil. This contamination can prove detrimental, leading to crop failures or reduced yields and harming young trees and other vegetation. Sulfur dioxide and nitrogen oxide particles present in the air can combine with water and oxygen to form acid rain. Primarily emitted by coal-fired power plants and motor vehicles, acid rain has far-reaching consequences. It alters soil composition, degrades water quality in rivers, lakes, and streams, damages crops, and accelerates the decay of buildings and monuments (National Geographic, 2023).

Carbon dioxide serves as a significant greenhouse gas, exerting a profound impact on global warming. It is released into the atmosphere primarily through the combustion of fossil fuels, including coal, gasoline, and natural gas. Modern society heavily relies on fossil fuels to power transportation, heat homes, and operate industrial facilities, consequently contributing to the emission of carbon dioxide. Additionally, other greenhouse gases emitted from both natural and human sources include methane, nitrous oxide, and fluorinated gases. Methane emissions are prevalent in coal plants and various agricultural processes. Nitrous oxide emissions stem from industrial operations, agricultural activities, and the combustion of fossil fuels in vehicles. Fluorinated gases, such as hydrofluorocarbons, are generated by industrial processes and are commonly used as alternatives to chlorofluorocarbons (CFCs), which have been phased out in many regions due to their role in depleting the ozone layer (National Geographic, 2023).

Globally, numerous countries have implemented measures to mitigate greenhouse gas emissions in order to address the challenges of global warming. One prominent example is the Kyoto Protocol, initially ratified in Kyoto, Japan, in 1997. This international agreement involves 183 countries committed to reducing their carbon dioxide emissions and collectively combating climate change.

## Indonesia

On August 31, 2023, Jakarta – the capital city of Indonesia – residents woke up to a thick blanket of haze and news that the city was ranked the most polluted city in the world by Swiss technology company IQAir. Air pollution spikes in the months of June, July, and August are a regular feature of life in Jakarta, occurring annually with the onset of the dry season, following monsoonal patterns. This is when a lack of moisture in the atmosphere and other meteorological conditions combine with high emissions, mostly from the transport and industry sectors, to produce dangerously high levels of pollution. In 2023, El Niño and ever-rising human emissions have created an especially dangerous spike in pollution.

At least there are five things to know about air pollution in Jakarta, from its natural and manmade causes, to its health impacts, to what can be done now to reduce the number of bad air quality days and lessen the burden of pollution on Jakarta's more than 10 million residents. First one is weather patterns drive pollution spikes, secondly the El Niño that intensifies weather's impact on air pollution and thirdly models show dry season conditions and high emissions create pollution hot spots in Greater Jakarta and East Jawa. Fourthly, transportation is an important local source of Jakarta's pollution, added by fifthly that industry and power plants also contribute to poor air quality. It is clear that air pollution significantly impacts public health in Jakarta, and Jakarta can and should move on solutions for now and the future (WRI Indonesia, 2023).

Indonesia also experienced forest fire from time to time. Study shows that there was a significant impact of haze to the human lung. There was a significant increase in respiratory conditions, lung function complaints and other related impacts. Further studies, especially cohort studies, should be undertaken so that the long-term impact of pollution from forest fires can be known (Aditama, 2000; Zaini et al., 2020).

## Policies Reducing Air Pollution

Reductions in air pollutants directly improve people's health in the near- and long-term. Policy actions to protect public health and well-being by addressing air pollution across sectors also positively impact the environment, economic prosperity, job creation, children's well-being, and gender equity, leaving an enduring legacy in countries and cities across the globe. Addressing air pollution by implementing well known and readily available solutions will improve health, drive sustainable economic development, and mitigate climate change. Polluted air is a collective problem; it knows no borders, and it requires collective action across nations and sectors to identify and implement evidence-based policies, innovative financing mechanisms and delivery models.

Addressing air pollution is an essential endeavor to safeguard public health on a global scale. Many sources of outdoor air pollution are beyond the influence of individuals, necessitating coordinated efforts by policymakers at the local, national, and regional levels. Collaboration across various sectors such as energy, transportation, waste management, urban planning, and agriculture is crucial to implement effective measures to curb air pollution and mitigate its adverse impacts. By working together, policymakers can enact policies and initiatives aimed at reducing emissions and promoting cleaner technologies to ensure cleaner, healthier air for all.

Implementing policies and making investments to support cleaner transportation, energy-efficient homes, sustainable power generation, cleaner industrial practices, and improved municipal waste management are crucial steps in reducing major sources of outdoor air pollution. Additionally, ensuring access to clean household energy can significantly alleviate ambient air pollution in various regions. Expanding the use of clean fuels and technologies is essential in reducing household air pollution and safeguarding public health. This entails promoting the adoption of solar energy, electricity, biogas, liquefied petroleum gas (LPG), natural gas, alcohol fuels, and biomass stoves that adhere to emission targets outlined in the WHO Guidelines (WHO, 2022).

Acknowledging the seriousness and immediacy of the issue, all member states of the WHO endorsed resolution A68.8, titled "Health and the Environment: addressing the health impact of air pollution," during the World Health Assembly in 2015 (WHO, 2022, December). This resolution was further bolstered by a comprehensive road map for action the subsequent year. In order to facilitate the reduction of air pollution levels and protect populations from associated health risks, WHO's Air Quality and Health Unit operates across three pivotal areas.

Firstly, it focuses on knowledge acquisition, evidence generation, and monitoring progress. Secondly, the unit is dedicated to enhancing institutional capacity and providing technical assistance. Lastly, it prioritizes leadership and coordination efforts to effectively address the challenges posed by air pollution on a global scale.

Beside policy and action in national and global level, anybody can take steps to reduce air pollution. Millions of people every day make simple changes in their lives to do this. Taking public transportation instead of driving a car, or riding a bike instead of traveling in carbon dioxide–emitting vehicles are a couple of ways to reduce air pollution. Avoiding aerosol cans, recycling yard trimmings instead of burning them, and not smoking cigarettes are others.

As a closing note, air pollution is the greatest environmental threat to public health globally and accounts for an estimated seven million premature deaths every year (UNEP, 2023). Air pollution and climate change are closely linked as all major pollutants have an impact on the climate and most share common sources with greenhouse gases. Improving our air quality will bring health, development, and environmental benefits.

## References

Aditama, T. Y. (2000). Impact of haze from forest fire to respiratory health: Indonesian experience. *Respirology*, *5*(2), 169–174.

National Geographic. (2023, December). Air pollution. https://education.nationalgeographic.org/resource/air-pollution/

UNEP. (2023, September 6). Air pollution note. https://www.unep.org/interactives/air-pollution-note/

WHO. (2022, December). Ambient (outdoor) air pollution. https://www.who.int/news-room/fact-sheets/detail/ambient-(outdoor)-air-quality-and-health?gad_source=1&gclid=CjwKCAiA_tuuBhAUEiwAvxkgTi8Rb9AIyVqrZcFQgpZzzlBaB9A6mV1Lgn8zm9THvvjFduZf8ReoBRoCKrMQAvD_BwE

WHO. (2023a, November). Air pollution. https://www.who.int/health-topics/air-pollution#tab=tab_1

WHO. (2023b, December). Household air pollution. https://www.who.int/news-room/fact-sheets/detail/household-air-pollution-and-health?gad_source=1&gclid=CjwKCAiAuYuvBhApEiwAzq_YiTM4pvU9bqt6yTdwydCabotY3gawXw2kZAOp3_x0k56PWzX2S76lBBoCmHMQAvD_BwE

World Health Organization. (2022, December 19). *Ambient (outdoor) air pollution*. World Health Organization. https://www.who.int/news-room/fact-sheets/detail/ambient-(outdoor)-air-quality-and-health

WRI Indonesia. (2023, September). 7 things to know about Jakarta's air pollution crisis. https://wri-indonesia.org/en/insights/7-things-know-about-jakartas-air-pollution-crisis

Zaini, J., Dwi Susanto, A., Samoedro, E., Bionika, V. C., & Antariksa, B. (2020). Health consequences of thick forest fire smoke to healthy residents in Riau, Indonesia: A cross-sectional study. *Medical Journal of Indonesia*, *29*(1), 58–63. https://doi.org/10.13181/mji.oa.204321

Chapter 2

# Vulnerabilities of Urban Communities to Climate Change

*Klaus Jurgens*

Economyfirst Limited London, UK

## Abstract

Climate change has become one of the "hottest picks" so to speak, not only in academic circles but in the world of mainstream media, too. Yet exactly a much more scientific approach is required to distinguish real concerns from sensationalist fake news which so often find their way into newspapers, radio, or television. Academia must refrain from copying non-fact-based (hypo-)thesis as climate change is too serious an issue as well. The approach of "build, shop, work differently" should lead to implementing the concept of self-contained and sustainable neighborhoods; reducing unnecessary daily individual traffic reduces avoidable pollution. We shall address the intertwined issue of whether or not elected office holders should propose such a change in consumer attitude, or if civil society itself would manage to first create the required awareness, then act accordingly. Introducing Slow Cities and Smart Cities leads to the major argument: not only working or shopping near to where we live is important but re-considering whether society should really depend on imported foodstuffs having been transported thousands of miles by air or sea, then lorry, or not better switch to a much more regional, environmentally friendly way of meeting ones' daily needs not to reject globalization per se but to create a regionalized, climate friendly variety of it. Finally, restructuring supply chain logistics might be a key development needed to successfully tackle and combat climate change on a global scale.

*Keywords*: Sustainable habitat; smart cities; slow cities; self-contained neighborhoods; supply chain logistics

Climate change has become one of the "hottest picks" so to speak, not only in academic circles but in the world of mainstream media as well. Yet exactly a

**The Social Consequences of Climate Change, 11–19**

**Published under exclusive licence by Emerald Publishing Limited**
**doi:10.1108/978-1-83797-677-520241002**

much more studied, scientific approach is required to distinguish real concerns from sensationalist fake news which so often find their way into newspapers, radio, or television. Academia – and by definition – must refrain from copying non-fact-based (hypo-)thesis as climate change is too serious an issue.

One solution to more logically approach the impacts of climate change on our daily life is to begin by analyzing whether the way we actually do go about that very daily life can be linked to the way we build our towns and cities. The argument put forward in this contribution centers around the personal assumption of the author that more sustainable neighborhoods are the key to a more ecological and environmentally friendly future. But what actually constitutes a sustainable neighborhood?

Let us begin with our very own cluster of blocks and streets; let us start by defining what actually constitutes a neighborhood and how big it could or should be before we embark on analyzing whether or not we need to enlarge the smallest units to eventually make the model of a sustainable neighborhood truly viable.

In "The drawbacks of a global concept of sustainable neighbourhood in developing countries" Mohamed Yazid Khemri, Silvio Caputo, Alessandro Melis argue that:

> A sustainable neighbourhood is generally considered the key spatial unit for cities, enabling sustainability to be applied as a set of urban planning, social and environmental principles (...) However, the concept of sustainable neighbourhood was developed in the global North, thus reflecting a city form and social context that do not apply to developing countries, which have different cultures, goals and priorities. Yet, in the global South, the uncontrolled urbanisation led to adopt the sustainable neighbourhood concept as developed for the cities of the global North, in an attempt to attain a more sustainable urban development (...) Ironically, this is inconsistent with the notion of sustainability that advocates a planning and design process customised to the local context. As a result, a rejection of the globalised projects by the local population for not being responsive to their lifestyle leads to spatial and social segregation (...) This paper aims to demonstrate the twofold hypothesis that the sustainable neighbourhood concept of the global North cannot be generalised, and a truly sustainable model should emerge from local context (Khemri et al., 2021).

Learning from Khemri, Caputo and Melis as quoted above we arrive at a first intermediate conclusion: if we agree that sustainable neighborhoods are indeed a possibility to redefine urban and rural living, we at the same time should agree upon that there is no one fits all – approach but that every country, every region in that particular country will need to design its very own model. Think mountainous or seaside regions, consider touristic or non-touristic towns and cities.

A related and extremely relevant issue is whether or not we can actually convince the citizens of those soon to be more sustainable towns and cities to accept that change is necessary. Change is never easy to be explained and often very cumbersome to be accepted by members of the public. Can the public be convinced? Well, as long as we understand that the public is a stakeholder group by definition and must be treated fairly and eye to eye, it may become a possibility. There is another policymaking domain in which that very public had been never truly consulted – international relations and foreign policies. We shall discuss the comparability between city planning and foreign relations by incorporating analysis from Christopher Hill as duly quoted by Michael and acknowledged in the paragraph below. Hence, let us now continue by focusing on public opinion as if we neglect it nothing what is proposed in this short article would ever really work out.

## Citizen's Acceptance or Rejection?

Should members of the public have a say, or at least interest in the foreign affairs of their country? Professor Christopher Hill, and it is only fair to say one of the leading scholars in the field of International Relations and in particular so during his long tenure at the London School of Economics and Political Science, clearly thinks so and published what can best be described a landmark book back in the year of 2003 (Hill, 2003).

In "The Changing Politics of Foreign Policy" Hill and in the words of Michalis S. Michael from La Trobe University wrote in December 2003 that:

> from the outset Hill is on a dual mission: to bridge the gap between the popular (or non-academic) perception of foreign policy and academic dismissal of foreign policy analysis as an archaic and irrelevant instrument of the state (p. 2); and to "liberate" foreign policy from its zealot custodians and "high priests" who since the eighteenth century have wrapped it in a veil of mystique "protecting" it from the realm of public policy and democratic accountability. Michael (2003)

Michael then continues by arguing "although to most people foreign policy is synonymous with international politics, it has become 'unfashionable' as a discourse of study, research and teaching amongst international relationists in academia" (Michael, 2003).

Then Michael adds his own scientific verdict:

> Systematic in his approach, Hill attempts to breathe new life into FPA, liberating it from the intellectual and inter-disciplinary constraints that have plagued it in the last few decades. He argues quite forcibly that foreign policy has been abandoned by IR and captured by historians (international or diplomatic history)

> and comparative analysts (country and area studies). In order to reinstate the relevance of international relations, Hill must also argue for the centrality, significance and validity of the state in world politics. Hill's strongest point is also his weakest. One area in foreign policy analysis that is the subject of renewed interest is the exploration of linkages with public opinion. In continuing this tradition, Hill calls for the inclusion of a democratic component in foreign policy formation in conjunction with efficiency and rationality, thereby endowing it with greater effectiveness and popular legitimacy. Michael (2003, pp. 101–109)

Assuming if explained correctly public opinion might be in favor we would then and in a next important step combine sustainable neighborhoods into a bigger picture so to speak – in other words, how will more sustainable neighborhoods help to create (more) sustainable entire towns and cities alike? Welcome to the concepts of Slow, and Smart Cities, respectively.

## Smart or Slow Cities or Perhaps Both?

The whole plan once put into action would, however, never result in any tangible benefits if we transform one local neighborhood into a "model sustainable neighbourhood" while 90% of our town or city remain untouched, stay put. In this further analytical step, we will introduce the master plan so to speak of recreating our habitat and not just planting a tree as laudable as it is in one chosen side street. Enter entirely sustainable towns and cities – smart, slow, or perhaps both?

From a methodological perspective, a number of inter-related impact assessments should be carried out as the author of this chapter suggested to a Turkish and European audience already back in the year 2006 (Jürgens, 2006). Incorporating the general relevance of the various types of impact assessment and analysis, respectively, and for our subject of sustainable neighborhoods as well as smart/slow cities we would best be starting by conducting a comprehensive assessment of the city's current infrastructure, systems, and services. This includes to identify key challenges and opportunities for improvement, such as traffic congestion, energy consumption, waste management, and public safety. A long-term strategic plan has to follow.

Once more returning to how Hill was referenced in Michael (see above) one of the most vital aspects in this regard is to engage with various stakeholders, including government agencies, local communities, businesses, academia, and technology providers and indeed to seek input and feedback from citizens to understand their needs, preferences, and priorities. In other words – to proactively engage with ordinary citizens during the planning stages and not simply by delivering end-results as well intended as they may be.

This approach includes infrastructure development including better access to transportation, promoting alternative means of commuting such as cycling where appropriate and most naturally the necessity to promote renewable energy sources such as solar, wind, and geothermal to reduce carbon emissions and dependence on fossil fuels.

But for a last time in this part of my contribution, let us state that long-term success of designing a smart town or city, respectively, depends to a large extent on citizen engagement and participation, must involve the collaboration with local communities and grassroots organizations to co-create solutions based on local needs and priorities.

Building a smart city is a complex and ongoing process that requires collaboration, innovation, and sustained commitment from government, businesses, communities, and other stakeholders.

In a world where speed is of the essence, where ever-increasing flows of individual traffic create havoc around town, where some workplaces declare that taking a break too long is considered cheating, where fast food has become the staple diet of the day so to speak and where students are confronted with an ever-increasing workload – at least our own four walls, our own neighborhood, our own town should provide us with some sort of peace and quiet, of well-being and ample opportunity for recharging our batteries. Thus said – welcome to the concept behind Slow Cities. Yet, that very concept is not about a withdrawal from our modern world, it is not about quitting your job to build a hut in the woods and take up hunting, it is not about an isolated single person or single-family lifestyle, respectively. It is meant as a revival of our local and regional communities under the banner of quality of life, it is aimed at having much more interaction with friends and neighbors alike yet without forsaking the comforts of information technology. It is geared at reconsidering whether that next prefabricated hamburger is the healthy option or whether we should better not frequent our corner restaurant and grocery stores instead where we know produce was locally sourced as well. It is about regional job creation as much as it is about attending the best universities two hours away without forsaking the comfort zone back home to be enjoyed at least every weekend. In this context let us quote from the official Slow Cities website and their chapter "About Us", wherein it is stated that "The Movement of cittaslow was born in 1999 through the Paolo Saturnini's brilliant intuition, past Mayor of Greve in Chianti, a little town of Tuscany. The new idea of considering the town itself and thinking of a different way of development, based on the improving of life quality, moved him to spread his thoughts all over our country." A major point we detect is that it was first and foremost about improving the quality of life for ordinary citizens and not just a lofty theoretical approach. What we thus understand is that it was about time not only to bring back the quality of life his town was once famous for but as a matter of fact increase it dramatically. Hence, we can argue that he was not against new developments but only if they would improve the local quality of life. In order to become a Slow City seven macro-categories are divided up into 72 subcategories ("How to Become"). Those macro requirements consist of stipulations for energy and environmental policies (think recycling); infrastructure policies (consider cycle paths); quality of urban life policies (for example fiber optics, wireless); agricultural, touristic and artisan policies (talking about prohibiting GMOs); policies for hospitality, awareness and training (the good welcome approach);

social cohesion (about prohibiting discrimination); and partnerships (for example cooperating with slow food promoters).

Now that we discussed the linkage between smaller units – sustainable neighborhoods – with entire towns or cities re-calibrating, one intertwined question we must address is whether or not we would risk to return to a semi-feudal style of living with landlords and peasants or whether or not we would actually enable our communities to prosper much more than they do now without pulling up the mental drawbridges? In a nutshell – would the concepts introduced in the previous chapters lead to anti-globalization, to building little local and regional fortresses and sarcastically written, returning to the notion of our earth is flat?

## Not Anti-globalization – Think Regional Equals Think Global 2.0

PIIE Peterson Institute for International Economics (Kolb, 2018) explains:

> Globalization is the word used to describe the growing interdependence of the world's economies, cultures, and populations, brought about by cross-border trade in goods and services, technology, and flows of investment, people, and information. Countries have built economic partnerships to facilitate these movements over many centuries. But the term gained popularity after the Cold War in the early 1990s, as these cooperative arrangements shaped modern everyday life.

The wide-ranging effects of globalization are complex and politically charged. As with major technological advances, globalization benefits society as a whole, while harming certain groups. Understanding the relative costs and benefits can pave the way for alleviating problems while sustaining the wider payoffs.

In a final step we shall now find out whether implementing sustainable ways of managing neighborhoods, towns, and cities is just a matter of improving our daily quality of life, the satisfaction and/or happiness factors, respectively, or whether changes in the way we live actually reduce carbon footprints, help to combat climate change and ultimately result in a greener, cleaner, much more ecologically motivated way of life? Are there, or are there perhaps no tangible benefits and if there are, who benefits the most?

Impact on supply chains and logistics and hence on tackling climate change.

We began this contribution by explaining the advantages and disadvantages of sustainable neighborhoods and whether the wider public is mature enough to fully grasp the entire dimension which would result in quite a number of serious changes in how we lead our everyday lives. We then introduced the concepts of both smart and slow cities, respectively; in all likelihood a combination will be the most logic solution. A closer look at which form of globalization, i.e., to which degree should our towns and cities be globalized, followed suit.

One aspect we did not as of yet analyze in detail is the expected impact on reducing global warming and thus more successfully tackling climate change. Let us hence assume the wider public and their elected office holders are satisfied with changes in and to their lifestyles and embrace the concept of sustainable neighborhoods based on a combination of the concepts of slow and smart cities and villages and towns. The key question for this book chapter would still need to be answered: will all this just result in a more "happy" population, will the population's quality of life improve and that was it, or are there tangible benefits for not just one region or country but the global society as a whole, benefits as in significantly reducing air pollution, reducing carbon footprints and in general supporting a more clean, more green, and as we mentioned in the first few lines of this article, a more sustainable way of life? According to the author of this article there are, and there are quite significant and highly tangible benefits.

Three examples:

First, a more regional form of managing our towns and cities promotes local and regional entrepreneurs and farmers who then more favorably compete with national or even multinational actors in the same domain. Think locally sourced tomatoes, consider regionally raised cattle. This in turn will create, or at least keep employment opportunities closer to where citizens actually live. Returning to the concept of Slow Cities, the food dimension is one of the major agenda points with a view to how it all started as we learned – locally or regionally produced and sourced foodstuffs, either to be consumed at home or in family (not anonymous chains!) run restaurants and boutique hotels and prepared according to either traditional, or nouvelle cuisine recipes as both are possible.

Second, as locally produced foodstuffs need to be sold in the marketplace, a clever supply and demand analysis must be in place, call it the sustainable neighborhood food supply chain business plan. One of its key advantages is the enormous reduction in land or air transportation which in turn greatly reduces our carbon footprint so to speak and thus helps to tackle climate change.

There will always be needs for transportation and there must always be enough of a profit margin for entrepreneurs and farmers as if not the public perception of the above mentioned changes in everyday life would of course be rejected at once. It must be a clever mix between producing and selling only regionally, or by splitting ones' output earmarking a certain percentage for being sold further away. It is for sure not an easy task but one that becomes a crucial element of successfully implementing all changes. Which brings us to the final point:

Third, no one wishes to deny a resident in a town 500 km away to enjoy foodstuffs or other products made exactly in that other location. Now another important factor arises – there is no need for thousands of large vehicles to transport goods each for their own client. The trick is to transport your produce to a reasonably distant logistics center by means of smaller

> sized vehicles as you would deliver less than a big full lorry and only then transfer them to big lorries indeed which would then bring them to another logistics center let us say 300 km away from the initial one. At the other end once more smaller vehicles pick up their loads and directly transport them to the end user, or the local distributors such as shops and supermarkets.

Even this example underlines that we very often should look back into the past of how goods reached their end-user. Think a farmer producing fresh milk. Every morning the milk he needs or wishes to be sold is placed outside his farm's entrance and a van or small lorry collects it all over the place, perhaps from 20 or 30 different farmers. Only then is it delivered to a cooperative or other holdings place from where it is either directly sold to markets and distributors, or transported further to a larger selling point including other forms of using milk in further products, think yoghurt or cheese.

This system worked perfectly and in many parts of Europe, it still does. Why not copy this approach and try and match it with most other foodstuffs or even manufactured goods as only two examples?

Tangible climate change benefits? Reduced number of vans and smaller lorries on our roads. Those that are required travel shorter distances. Logistics centers come closer to a production site or farm where you actually live, then transferring their loads to another logistics hub still reasonably close to the region where their goods would be finally sold. To facilitate a reduction in climate change and carbon footprints even more a combination of road and rail travel in-between two or more bigger logistics centers should become standard.

If we then incorporate our previously mentioned concerns about sourcing fresh – or frozen meat initially – from thousands of miles away instead of investing into the local farming community to only give one last example we further and drastically reduce carbon footprints. Again it must be part of a fair and logic profit margin plan as farmers in, for example, South America up to a certain extent to being allowed to export their foodstuffs to countries in North America or in particular Europe.

Supply and demand and then supply chain logistics will become some of the keywords of and for our sustainable future. While not taking away access to profit margins in faraway and/or developing countries first and foremost we should create local demand and at once local supply. With the above mentioned concepts of sustainable living in general, smart and slow towns and cities and a change in attitude among the wider public it may become truly possible to enjoy a healthier, cleaner, greener way of life. This concept is never against the notion of us living in an intertwined, interconnected world, exactly the opposite. What this concept aims at is by promoting a more regional approach to doing things and not just at home but in as many places as possible we help to rescue our planet for future generations, at home and abroad.

## References

"About Us". Cittaslow International. https://www.cittaslow.org/node/245

Hill, C. *The changing politics of foreign policy* (Palgrave Macmillan, 2003), here referenced to and by Michalis S. Michael, La Trobe University, December 2003.

"How to Become". *Cittaslow International.* https://www.cittaslow.org/content/how-become

Jürgens, K. (2006, July–September). Why regulatory impact analysis can overcome obstacles Turkey faces on her way to Europe? *Insight Turkey*, *8*, 3, 121–125 (Published by SETA VAKFI).

Khemri, M. Y., Caputo, S., & Melis, A. (2021, January). The drawbacks of a global concept of sustainable neighbourhood in developing countries. https://pure.port.ac.uk/ws/portalfiles/portal/25976509/A_global_concept_of_sustainable_neighbourhood.pdf

Kolb, M. (2018, 29 October). *What is globalization? And how has the global economy shaped the United States?* Peterson Institute for International Economics (PIIE). https://www.piie.com/microsites/globalization/what-is-globalization. Accessed on October 24, 2022.

Part 2

# Can Climate Change Be Considered as a "Social" Problem?

Chapter 3

# Climate Journalism Worldwide and in Türkiye: Anadolu's Greenline Project

*Serdar Karagöz*

Anadolu Ajans, Turkey

## Abstract

This article explores the impact of global warming and the climate problem by focusing on their reflections in the media. This chapter addresses climate journalism in line with Türkiye's domestic and international initiatives on climate and environmental concerns including the Zero-Waste Project and Anadolu Ajansı's attempts to change the narrative in the media about climate change via establishing a new editorial unit, Greenline.

*Keywords*: Climate change; 1.5 degree; climate journalism; Türkiye; Anadolu Ajansı; Greenline

"Men argue. Nature acts" – Voltaire Hanover College Department of History (1995)

## Introduction

The global community is currently grappling with the escalating issue of climate change, a phenomenon that has significantly intensified in its impact over the past three decades. This crisis presents a formidable threat to individuals worldwide, regardless of their financial status, age, or gender, and offers no discernible benefits. Addressing this crisis effectively necessitates a resolute and concerted effort from all sectors of society.

In this endeavor, various entities including governments, local authorities, civil society organizations, and the media play indispensable roles. The media, in particular, bears significant responsibility in this struggle. Its primary obligations include enhancing awareness among those uninformed about the climate crisis,

**The Social Consequences of Climate Change, 23–39**

**Published under exclusive licence by Emerald Publishing Limited**
**doi:10.1108/978-1-83797-677-520241003**

disseminating scientific evidence to counter the claims of climate change deniers, fostering cooperation among stakeholders engaged in addressing the crisis, and advocating for environmentally friendly practices within financial sectors.

In fulfilling these responsibilities, it is imperative for the media to adhere strictly to the fundamental principles of journalism and to adopt a strategy that prioritizes public welfare above all other considerations. By upholding these standards, the media can contribute significantly to global efforts to mitigate the impacts of climate change and safeguard the well-being of present and future generations.

In recent years, the heightened visibility of the ramifications of the climate crisis has facilitated the journalistic endeavors. However, the ubiquitous occurrence of these impacts across various regions on an almost daily basis expands the breadth of subjects and advancements that climate journalism, as a relatively emergent discipline, has yet to encompass. These diverse topics can be delineated into primary categories including greenhouse gas emissions, droughts, extreme weather phenomena, biodiversity depletion, forest conflagrations, plastic contamination, environmental degradation, sustainable development, and renewable energy initiatives. The inception of this thematic domain can be traced back to the Industrial Revolution.

The article will examine global warming, the climate crisis, and their effects on both everyday life and media, which have emerged as the most pivotal concerns for our planet since the Industrial Revolution. It will present insights into climate journalism, Türkiye's domestic and international efforts concerning climate and environmental issues, including the Zero-Waste Project, and Anadolu's efforts to reshape the narrative of climate crisis with Greenline and Environment Forum.

## The Impact of the Industrial Revolution on Climate

Since the onset of the Industrial Revolution to the present era, human activities have inflicted significant harm upon the Earth. Fundamental necessities such as sustenance and shelter, vital for human survival, have disrupted the natural equilibrium of the planet. These disruptions have led to the degradation of natural habitats, the extinction of species, imbalances in biodiversity, depletion of natural resources, alterations in seasonal patterns, occurrences of droughts, food scarcity, and other adverse consequences.

The expansion of urban centers and population growth spurred by industrialization, coupled with heightened emissions of toxic gases resulting from increased consumption, have contributed to the escalation of global temperatures. The primary drivers of anthropogenic climate change can be attributed to the widespread utilization of fossil fuels, namely coal, oil, and natural gas, which became prevalent during the Industrial Revolution. The combustion of these fuels releases greenhouse gases, commonly referred to as "carbon emissions," into the atmosphere. These gases act as a thermal insulating layer enveloping the Earth, trapping solar-derived heat and thereby causing a rise in global temperatures, commonly known as "global warming."

Additionally, the rapid melting of glaciers, vital in regulating the Earth's temperature and preventing overheating, is a direct consequence of this warming trend. Data collected from the European Space Agency's (ESA) decommissioned satellite CryoSat (Esa Climate Office, 2023), which monitored over 200,000 glaciers between 2010 and 2020, revealed that 2.72 billion tons of ice melted over the course of a decade. This loss accounts for approximately 2% of the world's total glacier mass. The disappearance of sea ice, which serves as a crucial buffer against the onset of an ice age, exacerbates extremes in the climate system and contributes to fluctuations in sea levels. As the planet continues to warm, it is experiencing significant shifts in climatic patterns, thereby exacerbating the threat of climate change.

## The Magnitude of Climate Change

To grasp the concept of climate change, or more accurately, the climate crisis, it is essential to delve into its definition. Climate change refers to prolonged alterations in the typical climate conditions expected at local, regional, or global levels. While this phenomenon encompasses the results of both human-induced activities and natural occurrences, it is notably evident that the primary driver originates from human activities spanning from the era of the Industrial Revolution to the present era. The Intergovernmental Panel on Climate Change (IPCC) (n.d.), headquartered in Switzerland and established in 1988 to evaluate the risks associated with climate change, employs temperature data from the period spanning 1850 to 1900 as a benchmark.

Climate change manifests numerous catastrophes beyond mere seasonal variations or occurrences of abnormally high temperatures. As outlined in the "2022 Drought Numbers" report compiled under the auspices of the United Nations Convention to Combat Desertification (UNCCD) (Yüksel, 2023a), droughts, desertification, wildfires, severe storms, and rising sea levels collectively impact an estimated 55 million people annually, representing the most frequently observed consequences. Nonetheless, owing to its contributions to the escalating temperatures of the Earth and oceans, as well as its role in the accelerated melting of glaciers, climate change poses a significant threat to the health and survival of numerous species. This underscores that the issue extends beyond mere meteorological anomalies, encompassing a profound threat to global biodiversity.

## Curtailing Global Temperature: 1.5°C Goal and Challenges

It is imperative to curtail the rise in the Earth's average temperature to 2 degrees Celsius, with further efforts aimed at constraining it to 1.5 degrees Celsius by the conclusion of the current century, relative to the pre-industrial era (1850–1900), to mitigate the catastrophic repercussions of climate change.

The 1.5-degree limit stands as a collective target established by 195 nations within the framework of the 2015 Paris Agreement, underscoring concerted global endeavors to mitigate climate change within this prescribed threshold (İklim Değişikliği Başkanlığı, n.d.). Nonetheless, prevailing progress suggests that attaining this objective presents considerable challenges. The commitments made by governments, institutions, and organizations thus far fall short of the necessary measures to confine climate change to 1.5 degrees Celsius.

The advancements realized through existing policies and pledges fail to meet the requisite standards. A more assertive approach to mitigation is indispensable to achieve the 1.5-degree target, necessitating a 42% reduction in emissions and a comprehensive transition to a carbon-neutral infrastructure by 2050. This entails a fundamental overhaul encompassing policies, energy systems, and economic frameworks. Despite scientific assertions that meeting this target may be unattainable, the absence of reinforced climate change policies portends a trajectory culminating in a 3.2-degree temperature escalation by century's end, as cautioned by U.N. Secretary-General Antonio Guterres, who warned of a perilous course leading to a 3-degree temperature rise.

Failure to attain the 1.5-degree target carries grave implications. Anticipated consequences encompass heightened frequency of scorching temperatures, increased fatalities due to extreme heat events, exacerbated and prolonged drought conditions, intensified and abrupt precipitation patterns, elevated sea levels, amplified occurrences of natural calamities, and biodiversity loss. These ramifications signify potentially calamitous impacts, including disruptions to global equilibrium and substantial disruptions to agricultural production.

Climate scientist Professor Dr Levent Kurnaz has forewarned that fatalities attributable to extreme heat could become commonplace, as highlighted in his assessment provided to Anadolu (Yüksel, 2023b).

No nation remains immune to the repercussions of climate change, primarily instigated by anthropogenic activities. Many realities have come to light and are gradually becoming manifest in daily life. Oceanic warming, biodiversity imperilment – with approximately one million species facing extinction – are among the pronounced concerns. Projections indicate that by 2050, at least one in four individuals will confront persistent water scarcity, while desertification and soil erosion exhibit escalating prevalence. Over the past three decades, forests equivalent to five times the expanse of Türkiye have vanished globally. Diseases stemming from air pollution, a foremost environmental menace impacting water quality and human health, account for nearly 10 million fatalities annually.

## The Climate Crisis Redefining Migration: Climate Migrants

Climate change transcends geographical boundaries and encompasses more than just distant phenomena like melting icebergs in Antarctica. It manifests as loss of life and land, devastating storms, floods, persistent shortages, and involuntary displacements.

All living entities – humans, flora, and fauna – are susceptible to migration pressures induced by climate change. Experts and climate models reveal that from 2010 to 2019, approximately 90% of displacement incidents were attributed to disaster-related extreme weather events such as floods, storms, and droughts. Moreover, projections suggest that by 2070, one in three individuals may be compelled to migrate due to climate-related factors. Consistent with numerous other challenges, reports from UN Women underscore the disproportionate impact of climate-induced migration on women and girls (Şeko, 2023a).

## Fundamentals of Combating Climate Change

Are we powerless in the face of the ongoing crisis, often characterized by the United Nations as a "ticking bomb," "climate destruction," or a "boiling age"? If we persist in our current habits, the answer to this question is an unequivocal yes.

During the record-breaking heat of 2023, the global daily temperature average exceeded the pre-industrial level by 1.5 degrees Celsius for a significant portion of the day (Erkul Kaya, 2024). This trend led to frequent headlines proclaiming "Heat Record Broken" and "A New Heat Record" across various news outlets.

While daunting, the gradual reduction or complete elimination of fossil fuels is within reach, offering a pathway to swift yet sustainable mitigation efforts.

Inger Andersen, executive director of the UN Environment Program (UNEP), highlights that the G20 nations, primarily responsible for 76% of all emissions, bear the primary responsibility to phase down and phase out fossil fuel consumption, thereby decarbonizing their economies (Erkul Kaya, 2023a). This underscores the core source of the crisis.

While the Global North exacerbates the climate crisis, the burden falls disproportionately on the Global South, encompassing South America, Africa, the Middle East, and South Asian countries, despite their minimal contributions to the crisis.

Ending this injustice is imperative to minimize the crisis's impact.

One proactive step involves maximizing the utilization of renewable energy sources, such as solar and wind power, generously provided by nature. These sources, including solar panels and wind turbines, represent the most cost-effective means to limit global warming to 1.5 degrees Celsius and mitigate emissions. Currently, renewable energy fulfills 30% of the world's electricity demands, a trend that must continue.

Transitioning to clean energy sources within public transportation, such as electric vehicles and bicycles, is critical to reducing emissions from conventional motor vehicles, significant contributors to air pollution and climate change. While governments, institutions, and companies play pivotal roles in combating climate change, citizens also bear crucial responsibilities. The top 10% of households with the highest per capita emissions contribute nearly half of greenhouse gas emissions from household consumption.

Individual actions, such as adopting climate-friendly practices in kitchens, waste reduction, proper waste disposal, recycling, and embracing environmentally friendly urban lifestyles, can significantly reduce emissions from household activities by 40–70% by 2050. Given that food-related emissions account for nearly one-third of global greenhouse gas emissions (ECBPI, 2021) and 60% of biodiversity loss, transitioning to a plant-based diet can have a substantial impact.

Simple measures to reduce energy consumption include switching to energy-efficient light bulbs, minimizing air conditioner and heater usage, water conservation, supporting afforestation efforts, and opting for plant-based diets.

While governments, companies, scientists, academics, NGOs, and environmental activists are often seen as key actors in addressing climate and environmental issues, the media also plays a crucial role. Half of the population stays informed about climate updates through television channels, news websites, and social media platforms.

When analyzing the influence of news agencies and broadcasting organizations in shaping public discourse, it is crucial to recognize that their interaction with the environment plays a significant role in combating climate change, particularly in engaging with readers or audiences.

## Specializing in Climate Journalism

At this juncture, we encounter the term "thematic journalism." In essence, it does not differ from basic journalistic practice. Although, it is a matter that both extends and confines the scope. Climate journalism, which focuses on climate and environmental issues, can also be assessed within this framework. It is a distinct specialization within science journalism, drawing its inspiration from all aspects related to climate. This field explores not just the core tenets of journalism but also terminology, the significance of the climate crisis, its impacts, and potential solutions, along with the reporting on these issues. While undertaking this task, it's crucial to gather information effectively and stay informed about current issues. This is a journalism that endeavor counters a constant flow, rapid changes and hollow concepts like "sustainability" and "awareness," which suffer from frequent but careless usage.

Full knowledge of climate terminology is one of the features that strengthen the hand of climate journalist. This maintains the journalists' resilience as they engage with scientists, companies, society, and individuals at the extremes who respond to climate change, whether positively or negatively. Approaching the right issues with the appropriate language, being sensitive to terminology, analyzing the obtained information, and eliminations are other steps in this process. A key aspect to highlight here is the bridge-building role between scientists and the public. This intermediary task, which aims to simplify complexity in a way, places a delicate responsibility on the media. Fulfilling this responsibility undoubtedly requires not only having journalists and media outlets well-versed or specialized in the subject matter, but also the ability to carry out the processes with utmost clarity and comprehensibility. This is exactly what journalism aims to do within specialized topics: You need to explain scientific subjects in a way that's brief, clear, and easily understandable to responding public queries.

Especially when it comes to digital publishing, such as websites, social media, and podcasts which are increasingly prevalent compared to conventional media like printed newspapers, radio, and even television, things become more challenging due to managing the uncontrollable speed. As scientific endeavors, climate journalism cannot afford to prioritize speed. Given that this field encompasses all lives, there is a matter of journalistic practice that requires extra sensitivity and a deliberate approach.

Sometimes, it may not just take days, but months to keep up with both current and long-term developments, explaining updates by backing them with expert reviews and scientific reports, simplifying scientific statements and reports into a more digestible and meaningful format.

This perspective actually challenges the norms of traditional journalism. The media generally prioritizes speed and staying current. This is valid not only for Türkiye but also for the world. While the social media gaining ground, news moving faster than ever, a shift toward broad strokes instead of finer details, the emphasis on disaster and catastrophe news that attracts more attention as an editorial choices, the rudity of the language used, the spread of disinformation at a terrible speed, the extremely weak verification mechanisms, the decrease in trust in the media, and people favoring "stories" over solid scientific makes pretty clear that things have become more difficult lately. Even with insufficient resources to address the climate crisis, which poses significant environmental and public health challenges, the media still plays a crucial role in guiding and influencing behavior. Journalism, inherently serving the public interest, bears an even greater responsibility when it comes to the health of species.

## The Importance of Climate Journalism

The climate crisis and environmental concerns have garnered significant attention in the media, with a notable surge in coverage observed since the 2000s, following earlier extensions into the public sphere during the 1970s and 1980s. While natural disasters have been a focal point of intensified interest, this heightened media attention has also been reflected in the film and documentary industries. However, despite the increased media interest, societies have not uniformly achieved a high level of awareness or exerted substantial pressure on responsible entities. Many countries still lack adequate information on the subject, with instances of information denial exacerbating the challenge of addressing the climate crisis. Moreover, persuading leaders who hold significant influence over a country's fate can be particularly challenging. Therefore, it is imperative for the media to recalibrate, regulate information dissemination, and enlist expert journalists to effectively cover climate journalism. The current global trajectory further underscores the urgency of this necessity.

The attention of the Turkish media has been increasingly drawn to climate journalism due to the vulnerability of Türkiye's geographical location. As a country heavily affected by the climate crisis, Türkiye experiences rising temperatures, drought, ecosystem degradation on land and sea, loss of wetlands and forests to wildfires, desertification, and pollution. These ongoing trends not only jeopardize human health but also pose a threat to the population at large.

Climate journalism encompasses a broad spectrum of effects across various issues, making categorization challenging. Given the global nature of the crisis, events in one region can have ripple effects elsewhere. For example, a melting glacier in Antarctica can contribute to rising sea levels in the Mediterranean. Journalists covering climate-related issues must navigate multifaceted socioeconomic contexts, impacting areas such as the economy, energy agenda, urbanization, and tourism. In such scenarios, reporters may face dilemmas regarding the prioritization of financial or political aspects over the core issue. Achieving balance represents a crucial yet challenging aspect of thematic journalism at this juncture.

## Sensitivity to Language in Reporting on the Climate Crisis and Natural Disasters

Language plays a critical role in climate news reporting, carrying inherent risks if not wielded with care and precision. Despite established standards in international journalism, reporting on climate and health topics demands a heightened level of responsibility due to the potential impact on public perception and behavior. Alarmist language, particularly in situations affecting communities and ecosystems, can perpetuate negative narratives and undermine the efficacy of informed decision-making.

"*The media is very willing to publish negative news about climate change*" stated Petteri Taalas, Secretary General of the World Meteorological Organization (WMO), to Anadolu (Arslan, 2023) in September 2023. Taalas points out that one of the challenges they face regarding climate change is the attitude of the media and press likes "scary and sharp stories" about climate change.

"*Whenever there is something negative about the climate, the media is very eager to publish it. Sometimes, even if the situation is not so dire, you get this impression when you look at the media outlets*" said Taalas and also emphasized that the negative news about climate change published by the media particularly scares young people, and he believes this is unfair to them (Arslan, 2023).

The case in point, in other words, the attitude of "disaster journalism" or, with a more precise definition, "doomsaying" overshadows the possibility of a solution and causes fluctuations by relegating proposed solutions to the background. It seems that there is no possibility to get progress without discussing what this stance might cost. Because this situation could result in precisely this: "(In case the existence of the climate crisis is accepted) *I can't avoid the climate issue; I'm facing a catastrophe with no solution in sight*." In a contrast situation, when the climate crisis is denied, it is possible to talk about a process that leads to a tendency to ignore it altogether, especially with the backing of media power and influence. Groups that stay hesitant, seeking guidance, or indifferent might find themselves pulled into a cycle of "insensitivity" due to the constant exposure to certain narratives. The course of this cycle is can fueled by a situation where media contributes to instilling feelings of hopelessness, perpetuating deadlock, and exacerbating disaster.

However, as previously mentioned during the definition phase, the source of the climate crisis lies in human actions and activities. This emphasis holds vital

Chapter 4

# Newton's Cradle Metaphor: Alternative Approach in Climate Change Education

*Şefika Şule Erçetin and Büşra Sarı*

Hacettepe University, Turkey

## Abstract

In this study, a metaphor showing that climate change education operates in dynamic interaction is presented. The metaphor allows us to see the broad context and importance of the effects of climate change education. In this study, it is aimed to raise awareness about the issue by drawing attention to climate change education with a metaphor. In this direction, theory and application were made by adapting Newton's cradle to the field of climate change education. If the impacts of climate change education are conserved as a whole through understanding and development, by staying in harmony, energy is not wasted and is conserved rather than misdirected. Increasing awareness and attention to climate change education, and adopting a broad-based approach to improving the system, are thought to be more effective and beneficial in shaping the future of climate change education and therefore combating and adapting to climate change. Newton's cradle metaphor has been a helpful tool to illustrate the dynamic interaction of climate change education. Perhaps this tool can help climate warriors who want to shape and steer the future to think creatively, gain perspective and see different potential solutions.

*Keywords*: Climate change education; Newton's cradle; metaphor; climate; climate change

## Introduction

It is not entirely possible to foresee the future of climate change education and to know what threats or opportunities it may face in the future. In addition, past data help us to understand the present, and some clues can be provided about

**The Social Consequences of Climate Change, 41–49**

**Published under exclusive licence by Emerald Publishing Limited**
**doi:10.1108/978-1-83797-677-520241004**

what may happen in the future and how it may happen with existing premises. With a well-prepared climate change education, predicted problems can be faced in the future. Although climate change education has recently gained awareness among scientists and policymakers (Barak, 2020), the impact of this education on combating climate change is not clear. For this reason, this metaphor study was conducted to draw attention to the effects of climate change education. In this way, it is aimed to raise awareness about climate change education.

On the other hand, while the increasing international effort to address climate change has contributed to the increase of scientific literature on climate change action, there is a little discussion focused on climate change education (Singh, 2020). It is thought that this study will contribute to these discussions. In this direction, after explaining Newton's Cradle model, which is the subject of metaphor in the study, this model was adapted to climate change education. The purpose of using Newton's cradle model is not only to add a dynamic feature to the study, but also to support the assumption that climate change education has a significant impact, and to demonstrate the interaction of existing forces in education by embodying it.

### *Newton's Cradle Model*

Newton's cradle is a simple device that elegantly demonstrates the conservation of momentum and energy, the most fundamental laws of physics (Hutzler et al., 2004; Schulz, 2012; What is Newton's cradle?, 2022). It is also called by different names such as Newton's pendulum, Newton's balls, Newton's rocker, or executive ball clicker (Wikipedia, 2022). It is sold as an executive table toy and is used in physics textbooks to show the operation of the principles of conservation of momentum and kinetic energy in elastic collisions (Gauld, 2006).

There is much confusion about the origins of Newton's cradle. The physics principles used in the cradle, which is thought to have been invented by Newton based on its name, were actually put forward long before Newton's physics career. The 17th-century French physicist Abbe Mariotte was the first to demonstrate the law of effect. In 1662, John Wallis, Christopher Wren, and Christiaan Huygens presented papers explaining theoretical principles. Sir Isaac Newton also acknowledged Mariotte's work as pioneering work with Wren and Huygens in his 1687 book Philosophiæ Naturalis Principia Mathematica (Mathematical Principles of Natural Philosophy) (Harvard Natural Sciences Lecture Demonstrations, n.d.; Schulz, 2012; Wikipedia, 2022). The current name of the device was given by the famous actor Simon Prebble in 1967. The name gained acceptance because Newton had a much greater overall impact on physics. Prebble originally designed the cradle with a wooden frame and steel balls to market in-house. Later a stylish chrome frame became a common design (Hutzler et al., 2004; SAE International, n.d.; What is Newton's cradle?, 2022; Wikipedia, 2022).

Generally, five or seven balls (can be any number) suspended on two parallel bars, barely touching each other and suspended in the same plane, are used to demonstrate the laws of physics in the setup (see Fig. 4.1). The use of parallel bars

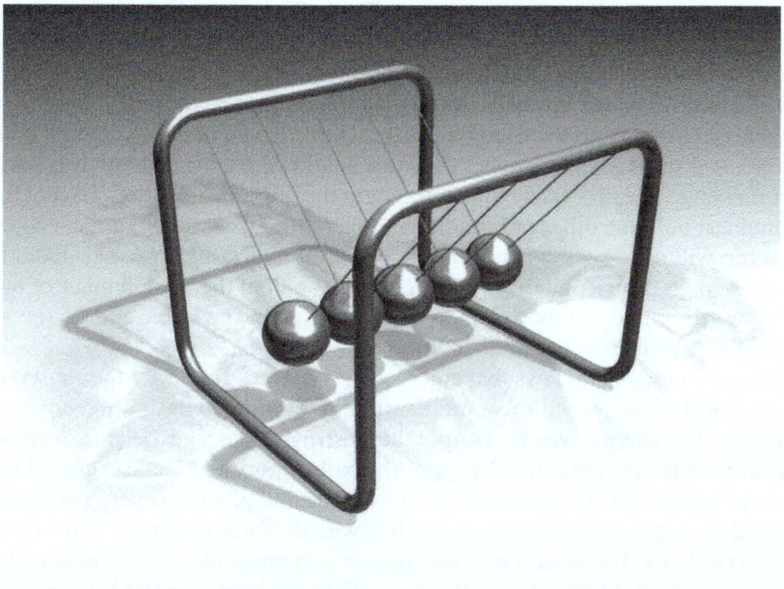

Fig. 4.1. Newton's Cradle Model (Wikipedia, 2022). This photo is licensed under Unknown Author, CC BY-SA-NC.

allows the balls to swing in a single plane. If the balls could move in different planes, the apparatus could not be used for the correct representation of the laws. Also, all balls should ideally be the same size, weight, and density. It may work for balls of different structures, but the clarity of the representation of the laws of physics will be difficult in such a case. Balls are specially made of stainless steel balls because steel balls are quite elastic, contrary to popular belief. A very elastic and uniform-density material is needed for the movement of the cradle, and as the elasticity of the balls increases, the duration of their movement will increase. On the other hand, the density of the balls must be the same to ensure as much energy transfer as possible. If the material density of the balls is changed, the way of energy transfers between the balls will also change. However, as long as the balls are all the same size and density, being too big or too small will not change the result (Harvard Natural Sciences Lecture Demonstrations, n.d.; Schulz, 2012; What is Newton's cradle?, 2022). The conditions necessary for the apparatus to work in the representation of the laws of physics can be summarized as follows:

- Balls should be aligned in a single plane as best as possible.
- The balls should hang in such a way that they are barely touching each other.
- The balls must be the same size.
- The densities of the balls must be the same.
- Balls must be elastic.

Newton's cradle has a very simple use. The ball at one end of the string is pulled apart from the others and dropped from a certain height. When this first ball falls, it hits other stationary balls, and a force is transmitted through these balls that push the last ball at the other end upwards. As a result of this impact, the ball at the end of the string moves away from the fixed balls to reach the height at which the first ball was released, and stops when it reaches the same height where the first ball was released. After this stage, the process begins to repeat in the opposite direction. Except for the first and last ball, the other balls in the middle of the string do not appear to move. The setup will also work when multiple balls are pulled and released. Even if the number of balls drawn and dropped is more than half of the total number of balls in the string, the same number of balls will move from the other end of the string (Ahmad & Anwar, 2019; Gauld, 2006).

The two fundamental laws that emerged in Newton's cradle are the conservation of energy and the conservation of momentum. According to the law of conservation of energy, the energy created cannot be destroyed but can be transformed. In Newton's cradle, the potential energy of the balls turns into kinetic energy. The first ball stops when it hits the second ball. The kinetic and potential energy of the ball is reset and the energy passes to the second ball with the force of impact. This continues between the balls. The reason why the motion is in one direction is explained by the law of momentum. Momentum is the force of objects in motion (SAE International, n.d.). Momentum is conserved just like energy, and specifying the direction of the force is part of its definition. When the first ball hits the second ball in the cradle, it determines the direction of motion, hence the direction of momentum. Any change in the direction of motion – but with the effect of an external force – also changes the momentum. However, since the cradle is not a closed system, a situation arises where the conservation of momentum is violated. When the last ball reaches the top, it comes to a short but definite stopping point and remains motionless, so momentum cannot be conserved. This is due to the existence of an external force, namely gravity. Another factor affecting the operation of the mechanism is elastic collisions and friction. With elastic collision, the balls stay the same before and after the collision and transfer the energy along the line. Since there is an elastic collision, the balls can continue to swing indefinitely, but after a while, they slow down. The reason for this is the existence of friction force. Friction is affected by the energy system, causing the balls to slowly come to a standstill. The source of the friction comes from a small amount of air resistance and mainly from the balls themselves. The balls cannot escape the effect of friction and yet their total energy remains the same due to energy conservation. Also, the vibration of the balls creates a clicking sound. Imperfections such as the balls not being perfectly aligned, and not having the same density, slow down the cradle. Energy deviations cause all balls to swing together after a while (Schulz, 2012).

Based on the approach that Newton's cradle model used in the study can have different meanings other than its intended use, and that this mechanism, which contains the laws of physics to understand and interpret what is happening in the environment, can be studied from a different perspective, in the next section, climate change education will be tried to be explained and interpreted with the metaphor of Newton's cradle.

## Climate Change Education and the Newton Model

Climate is defined as the average meteorological events seen in a place for a certain period (T.R. Ministry of Health, Public Health Institution of Turkey, 2015). Climate change, on the other hand, is the changes over time in weather statistics, which mean values differ from normal for various reasons (Şıvgın & Afacan, 2020). One of the important issues to be considered for combating and adapting to climate change is education. All kinds of education and training activities are applied to turn individuals into climate literate to ensure sustainability is defined as climate change education.

The metaphor of "Newton's cradle" will be used to understand the dynamics and uncertainties in climate change education. A metaphor is defined as a metaphorical tool that connects concepts in different ways, making the strange familiar and the familiar strange. Metaphor pushes the boundaries of traditional ways of thinking and creates the potential to rethink concepts with creative thinking. It can be used to discover new ways of thinking and produce solutions to open-ended and ambiguous issues (Lakoff, 1989; Phillips, 1998). Meanings can be embodied or suggested for those working in different disciplines. Thus, knowledge that includes unity and is shared between different groups of people can be created. In addition, metaphor can provide people with access to hidden nuances and complexities by giving a positive perspective to situations that seem negative (Mcallister et al., 2014).

Considered as a metaphor, we can position climate change education in the place of the first ball that started the movement in Newton's cradle mechanism. For the mechanism to work, for the transfer of energy and the conservation of momentum to be observed, and for a ball to change its place, this first ball must always be in motion and get rid of inertia. The movement of the first ball will be sufficient for the movement of the other balls (Özcan, 2017). It can be said that climate change education will make the first start that will accelerate to take action in the combat against climate change (Climate Change Education | UNESCO, n.d.). With climate change education, not only the education itself will be freed from inertia, but also the forces it will affect, just like other balls in the mechanism, will be able to save them from inertia. Just as energy cannot be destroyed and protected, this effect of climate change education cannot be destroyed and preserved. Therefore, the impact of climate change education on other dynamics is inevitable.

Climate change education and development have their strengths. Emergencies such as climate change have a collective impact and require solidarity (EKOIQ, 2022). Thinking metaphorically reveals the CCE's position in a complex web of long-standing issues. It also shows that this network creates fluctuations, each affecting the other, colliding, pulling in different directions, and causing energy. In the dynamic interaction of Newton's cradle, it is possible to imagine climate change education as balls in the cradle and to see each of them as important parts that affect each other, instead of trying to understand them separately (see Fig. 4.2).

In Newton's cradle, the balls had to be elastic for the continuation of their movement. The more flexible the balls are, the longer the movement would last. Similarly, if climate change education has a flexible structure, it can be said that

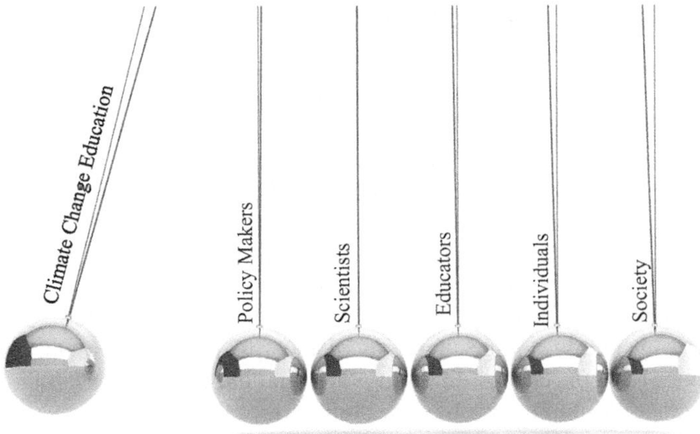


Fig. 4.2. Climate Change Education Dynamics. This photo is adapted by the authors from the Microsoft 365 creative content library.

its effects will last longer. What is meant by flexibility here is not only that the educational practice should be flexible, but also that it will be an education that covers individuals of all ages according to the principles of lifelong learning, repeated in a flexible time frame rather than a one-time education. When such education is provided, individuals will be able to gain the ability to combat climate change for a longer period.

For the mechanism to work properly in Newton's cradle, all balls in the mechanism must be aligned in a single plane. In this direction, all the activities, activities and studies carried out or planned for the sake of education in climate change education will continue in the same plane and direction that will give strength and energy to each other, and focus on a common goal, just as in the mechanism, will provide an accurate and realistic representation of the effects of climate change education. If one cannot move and focus on common ground, the movement of the balls will slow down after a while and the effects of climate change education will decrease. Energy deviations may also occur, causing the efforts made in the name of education not to have the desired effect and even the loss of the effect of education.

The balls hung on two parallel bars in Newton's cradle, barely touching each other, can be compared to institutions and organizations involved in climate change education at the national and international levels. It can be said that it is essential for education that they lead like two parallel bars. The balls hung so that they barely touch each other can be considered as the necessity of regional differences in the combat against climate change. This situation can be evaluated as it would be beneficial and correct to prepare and use training contents that are very close to each other but also differ according to regional needs.

For the effects of climate change education to be observed clearly, it can be said that the dynamics of society should give equal importance to this education.

This situation can be likened to the fact that the representation of the laws of physics cannot be clear when the sizes of the balls in the cradle are not the same. Similarly, it can be said that these dynamics should focus on climate change education with the same intensity. Otherwise, the effect of climate change education on dynamics may change, as the energy transfer pattern between balls with varying densities will change.

According to the amount of ball held and released in Newton's cradle, the amount of ball that is affected by this on the other side is the same. The participation rates in climate change education will be similar to this, and the increase in participants who will take action will be affected by climate change education and will increase in those who take action at the same rate. On the other hand, in the Newtonian cradle model, there will always be negative external factors that affect climate change education and reduce the effects of this education, such as the existence of friction and gravity forces that steal the balls from the energy system and slow them down and even cause them to stop after a while.

## Conclusion

Newton's Cradle model was used as theory and practice in this study, which was carried out to raise awareness of climate change education by drawing attention to it with metaphors. In this direction, after explaining Newton's Cradle model, which is the subject of the metaphor, the connection between climate change education, and the model was tried to be established.

With Newton's cradle metaphor, it becomes possible to see the broad context and importance of the effects of climate change education. If the forces within climate change education each maintain their integrity through understanding and development, by staying in harmony, energy is preserved rather than wasted or misdirected. The metaphor helps to explain the impact of climate change education on combating climate change. In this way, increased awareness and attention, a broad-based approach to systemic improvements will ensure that climate change education, and therefore climate change combat and adaptation, will be more energy efficient and progressive.

When the dynamics of climate change education are considered together with the relations they establish with each other, they can create a greater vision by interacting, connecting, or colliding from time to time, like the forces in Newton's cradle. It can be said that awareness of this great vision is the prerequisite for climate change education to have an impact on both education and the future of combating the climate crisis and building and controlling it. To employ all actors in the combat against climate change, it is necessary to put climate change education into action first. However, it is understood that the fact that the said actors act on common ground and focus on the subject with the same intensity can increase the impact of climate change education. Thanks to the metaphor, it has become clear that this training should have a flexible structure in terms of time, place, method of application, and participants.

In summary, Newton's cradle metaphor has been a helpful tool to illustrate the dynamic interaction of climate change education. Perhaps this tool can help climate warriors who want to shape and steer the future to think creatively, gain perspective, and see different potential solutions.

## References


Ahmad, M. S., & Anwar, M. S. (2019). *Newton's cradle* (ss. 1–5).

Barak, B. (2020). Dünyada ve Türkiye'de Ortaokul Öğretim Programlarının İklim Değişikliği Eğitimi Yaklaşımına Göre Karşılaştırılması. *Journal of Geography*. https://doi.org/10.26650/jgeog2019-0039

Climate Change Education | UNESCO. (n.d.). Tarihinde 30 Nisan 2022, adresinden erişildi. https://www.unesco.org/en/education/sustainable-development/climate-change

EKOIQ. (2022, Ocak 27). *COP26'da İklim Değişikliği Eğitimi Alanında Neler Yaşandı?* https://ekoiq.com/2022/01/27/cop26da-iklim-degisikligi-egitimi-alaninda-neler-yasandi/

Gauld, C. F. (2006). Newton's cradle in physics education. *Science and Education*, *15*(6), 597–617. https://doi.org/10.1007/s11191-005-4785-3

Harvard Natural Sciences Lecture Demonstrations. (n.d.). *Newton's cradles*. Harvard University. Tarihinde 13 Nisan 2022, adresinden erişildi. https://science demonstrations.fas.harvard.edu/presentations/newtons-cradle

Hutzler, S., Delaney, G., Weaire, D., & MacLeod, F. (2004). Rocking Newton's cradle. *American Journal of Physics*, *72*(12), 1508–1516. https://doi.org/10.1119/1.1783898

Lakoff, G. (1989). *Women, fire and dangerous things: What categories reveal about the mind*. Içinde The University of Chicago Press. https://doi.org/10.1111/j.1468-0017.1989.tb00245.x

Mcallister, M., Madsen, W., & Holmes, C. (2014). Newton's cradle: A metaphor to consider the flexibility, resistance and direction of nursing's future. *Nursing Inquiry*, *21*(2), 130–139. https://doi.org/10.1111/nin.12037

Newton'un beşiği nedir? (2022, Şubat 15). https://www.netinbag.com/tr/technology/what-is-newtons-cradle.html

Özcan, B. (2017, Şubat 12). *Newton'un Beşiği*. https://barisozcan.com/newtonun-besigi/

Phillips, B. (1998). Energy and performance: The power of metaphor. *Career Development International*, *3*(1), 18–22. https://doi.org/10.1108/13620439810368583

SAE International. (n.d.). *Newton's cradle*.

Schulz, C. (2012, Nisan 8). *How Newton's cradles work*. HowStuffWorks. https://science.howstuffworks.com/innovation/inventions/newtons-cradle.htm

Şıvgın, B., & Afacan, S. (2020). İklim Krizi ile Küresel ve Bireysel Mücadeleye Yönelik Bir Çalışma: Greta Thunberg Örneği. *Diplomasi ve Strateji Dergisi*, *1*(2), 75–92.

Singh, S. A. (2020). Climate change education: Knowing, doing and being. *International Research in Geographical & Environmental Education*, *29*(4), 362–365. https://doi.org/10.1080/10382046.2019.1703318

T.C. Sağlık Bakanlığı Türkiye Halk Sağlığı Kurumu. (2015). *İklim değişikliğinin sağlık üzerine olumsuz etkilerinin azaltılması ulusal programı ve eylem planı*. https://hsgm.saglik.gov.tr/depo/birimler/cevre-sagligi/2-ced/iklim-degisikligi/Iklim_Degisikligi_Eylem_Plani.pdf

Wikipedia. (2022, Ocak 22). *Newton's cradle*. https://en.wikipedia.org/wiki/Newton%27s_cradle

Chapter 5

# Risks or Challenges of Climate Change: The Case of Climate Refugees

*Brígida Brito*

Universidade Autónoma de Lisboa, Portugal

## Abstract

The causes of climate change and extreme events have been widely discussed in academic and scientific circles. Between natural and historical factors characterized by determinism and imponderability and the anthropogenic factors resulting from human, social, and economic-productive evolution, consensus is not always easy to define. However, scientific research has increasingly shown that human action has contributed to the great acceleration of the effects since the emergence of the era of humans.

More than just a list of causes, it is important to understand the impacts in different regions of the world, particularly in terms of the affectation on people. The populations that suffer the most from the effects of extreme weather events are the poorest, who live in close dependence on nature and therefore suffer from vulnerability without foreseeing the risks they face. The population that is most affected and often suffers from threats to their lives is forced to leave, either individually, in groups, or massively with no return date. Today, the international community recognizes the need to analyze and frame these people who seek international protection without being offered and guaranteed.

This chapter presents a discussion of the causes of the multiplication of extreme environmental events, particularly climatic events, in order to better understand the social and human implications of the production of forced social movements due to environmental causes. Although multilateral organizations recognize the existence of new categories of people in mobility for environmental reasons, there is still a gap in the framework of international law, which represents a global challenge.

**The Social Consequences of Climate Change, 51–65**

**Published under exclusive licence by Emerald Publishing Limited**
**doi:10.1108/978-1-83797-677-520241005**

*Keywords*: Anthropocene; climate change; risks; challenges; forced displacement; environmental refugees

## Introduction

The topic of climate change has been studied and analyzed for several decades, generating complex and not always consensual debates. As Edgar Morin says, a complex reality refers to "(...) the inability to be sure of everything, to formulate a law, to conceive of an absolute order (...) to realise that we can never get uncertainty and that we can never have total knowledge" (2007, p. 69).

After the environmental issue was consolidated in the 1970s, the concern on climate variations became more relevant, as a result of the realization of two very important facts: firstly, the global warming is a reality that poses risks and threats to territories and populations, and secondly, the evidence of the negative socio-environmental impacts that extreme weather events produce.

In the many revisionist scientific studies that have been conducted by the Academy, the concept of global warming is associated with the rise in the temperature of the air and marine waters, producing multiple and damaging socio-environmental effects, such as rising sea levels (Oceano Azul Foundation, 2021, p. 9).

However, with regard to the analysis of the impacts produced by climate change, there is a more comprehensive and complex process than the one identified for the phenomenon associated with global warming. In this sense, the emergence of extreme occurrences with increasing regularity is evidenced, but also of progressively greater intensity in terms of the effects generated (Santos, 2021, pp. 37–42). Currently, in-depth knowledge of the subject means that the events associated with global warming are one of the signs of climate change.

The relevance of the issue is evidenced by the factors that trigger the occurrences and pose a risk to the sustainability of the planet. Above all, by the threats that affect populations who, for a wide variety of reasons, live in a situation of fragility and closely dependent on nature. In the face of extreme weather events, vulnerable populations suffer from exposure and sensitivity (International Displacement Monitoring Center, 2015; White, 2011), and are affected by the impacts that are produced. In these circumstances, the management models for the damage caused are, in many contexts, ineffective and unresponsive, which means that vulnerability worsens. The impact tends to create and exacerbate uncertainty, which results in the forced displacement of vulnerable populations. These population movements occur mostly on a massive scale and due to environmental causes, although there is not necessarily any associated evidence of violence, persecution, genocide, conflict or war. They are primarily the result of climatic or geophysical events (IPCC, 2022; Maplecroft, 2017; Sui, 2022).

The whole problem of forced displacement for environmental reasons, whether internal or taking on an international dimension because of crossing borders, demonstrates what is known as the climate issue. In many cases it coincides with the occurrence of geophysical events, such as seismic activity, with or without the creation of tsunamis, and volcanic activity. On the one hand, extreme weather events represent multiple risks because jeopardize the achievement of sustainability, materializing into threats to human life. On the other hand, they are seen as emerging challenges that call for urgent intervention on the part of a multiplicity of actors, especially the state, global partnerships involving multilateral action, civil society organizations, and the affected population itself. The challenges to governance with the involvement of different and multiple stakeholders stand out, highlighting the fragility of conventional management models and calling for a reinvention of practices that can be considered responsible and, at international level, promote socio-environmental justice.

In order to analyze this topic, a methodology centered on bibliographical research was followed, particularly oriented toward conceptual and documentary definitions, guaranteeing the presentation of cases that exemplify the socio-environmental impact of extreme weather events with the creation or worsening of forced displacement due to environmental causes.

## The Climate Problem: From Historical Naturalism to the Anthropocene

The analysis of the climate issue has led to a scientific discussion, not always consensual, about the causes that promote the emergence and worsening of extreme events. An in-depth analysis of the different approaches identifies three positions, two of which are clearly distinct: the naturalist perspective with an evolutionary orientation; the anthropogenic approaches; and the mixed analyses.

Firstly, the so-called conventional naturalist interpretations present historical-evolutionary and natural factors as the main cause of climate change. The impacts are understood as a mere consequence that requires planetary adaptation, which eventually occurs through self-regulation mechanisms. The historical and evolutionary analysis is confirmed by the assumptions of Milankovich's analysis which, through an interpretation of the Earth's orbital parameters, explains the alternation of glaciations (Pultarova, 2022). Thus, the focus of the justification centers on the evolution realized by the transitions of glacial eras, seeking to demystify and deny human influence in creating or worsening changes in climatic conditions that can become unsustainable for human life. For the deniers – or skeptics – of human influence on the climate, the explanation focuses on the description of the historic transitions of ice ages that lead to decisive changes from an environmental point of view, extinguishing species, forcing some to adapt and even allowing others to emerge. The aspect that is not denied is the occurrence of transformations in ecosystems, even though these are natural processes resulting from an evolution that is impossible to reverse. Some authors can be identified with these interpretations marked by historical determinism, such as the meteorologist Piers Corbyn, the researcher Nils-Axel Mörner, the counselor Christopher Monckton and the geographer Antón Uriarte. For them, any reading that relates human influence with climate change is associated with an international lobby.

Secondly, as opposed to conventional naturalistic approaches, various studies by individual researchers, such as Paul Crutzen, Eugene F. Stoeme, Brian Fagan, Will Steffen, Pablo Meira Cartea, or Paulo Magalhães; by teams, such as the Stern Report; and by multilateral agencies, such as the Intergovernmental Panel on Climate Change (IPCC, 2018, 2019, 2022) or the United Nations agency World Meteorological Organisation (2023), indicate human factors as the main causes of climate change. Although they start from the realization that global warming is a reality, particularly in terms of increasing air and ocean temperatures and, consequently, rising sea levels. These readings focus on extreme weather events as a whole, defining them as holistic occurrences with a systemic scope. The factors that trigger the emergence or worsening of extreme weather events are the result of human activity, and are therefore referred to as anthropocentric.

Most of the authors who defend human influence on the climate define the industrial period as a milestone, an era that was profoundly transformative from an economic, social and environmental point of view. With the Industrial Revolution of the 18th century – and also the Transport Revolution – a new era began, marked by human activity that sustains itself on nature, overusing and degrading it, which is called the Anthropocene (Maldonado, 2018).

The Anthropocene concept, which defines the postindustrial period, has been widely analyzed by authors such as Paul Crutzen and Eugene F. Stoeme (2000), and Will Steffen et al. (2015). This is a period characterized by the transformation of industrial, transport and energy-based production, with a dependence on coal and its consequent burning, triggering previously unknown levels of greenhouse gas (GHG) emissions. From the moment that the economic sector became dependent on production and consumption models with the use of highly profitable energy resources (coal, oil, natural gas), the conditions were created for the emergence of the supermodernity characteristic of the Age of Man or Age of Humans (Crutzen & Stoemer, 2000; Crutzen, 2002; Steffen et al., 2015).

Will Steffen et al. (2015) define the earth system as fragile and vulnerable in the face of natural disasters, questioning the capacity for planetary self-regulation, suggested by naturalists, in the light of the human tendency to destroy nature. The definition of great acceleration (McNeill, 2000) is presented in terms of what Steffen et al. (2015) also refers to as complex processes, clearly characteristic of industrial and post-industrial societies that coincide with the Age of Humans framed by Crutzen and Stoemer (2000). These complex processes are twofold. On the one hand, at a global level, they center on continuous – and unstoppable – demographic growth, resulting in an increased and permanent burden on naturally vulnerable spaces and resources which, due to human action, suffer from exponential fragility. For Will Steffen et al. (2015), this is a problem that can be considered with a geographical dimension, given that the pressure exerted by population growth and density is particularly felt in certain regions of the world, particularly in southern countries. On the other hand, complexity is guided and defined by the propensity for exploiting natural resources and burning fossil fuels, since these represent the energy basis of industrialized societies. Great acceleration is therefore understood as a process of transformation of nature promoted or aggravated by man. The human drive to have more, in the sense of production and growth, which is characteristic of the capitalist economic model, is proving to be unstoppable and infinite, based on an energy pattern dependent on the exploitation and use of fossil fuels with an impact on an unprecedented scale (McNeill, 2000). In framing the concept of great acceleration, Will Steffen et al. (2015) present a set of socioeconomic indicators, including demographic growth, with emphasis on the urban population, real GNP, energy consumption, transport, foreign direct investment, tourism, access to and use of water. For all the indicators, it can be seen that the great acceleration began in the 1950s, with an intensity of growth associated with postindustrial progress.

According to these anthropogenic perspectives and based on the concept of the Anthropocene, industrialized societies are considered to be the ones causing a more profound worsening of climate variation around the world, contributing to the emergence and multiplication of extreme events with a socio-environmental impact. Energy, production, and consumption patterns are now seen as the main factors that accentuate the level of GHG emissions. Between 1990, the timeframe of reference for analyzing sustainability, but also climate, and 2020, GHG emissions increased by 41% (Ferreira, 2020, p. 39; Santos, 2021, p. 24). Of these, 73% concern to nonadapted energy consumption, including in the transport, electricity, heating, construction, building management, and manufacturing sectors. These data are concentrated in 10 of the world's richest countries or groups of countries with the highest levels of industrial development: China, the United States of America, the European Union, India, Russia, Japan, Brazil, Indonesia, Iran, and Canada (World Resources Institute, 2021).

The third approach to the causes of the emergence and worsening of climate change is moderate compared to the previous two, in the sense that it draws on relevant contributions from both. On the one hand, historical evolutionism is recognized as valid, based on the identification and characterization of ice ages that resulted in climatic transformations with a confirmed impact, particularly with regard to the structure of ecosystems and biodiversity. It is then accepted that the Earth has gone through periods marked by different climatic patterns, sometimes cooling and sometimes intensely warming, causing effects that can be defined as devastating whenever an ecosystem is irreversibly transformed or a species becomes extinct. But on the other hand, human influence on the climate is assumed, starting with industrial societies, the Era of Humans.

## Elements of Global Climate Risk Inequality

The year 2020 has been defined as the hottest year on record (Ferreira, 2020, p. 40). According to prospective analyses presented by the United Nations Environment Program (UNEP), if the trend continues until 2100, a third of fauna and flora species will become extinct. However, throughout the world, the human scenario appears to be more serious, as some territories are expected to face socio-environmental situations characterized by imbalance, namely more frequent prolonged droughts with a significant impact on local, national, regional and global food security. At the same time, prospective trend analyses suggest to a rise in sea levels of up to one meter, causing frequent floods with destructive effects, in some cases irreversible, leading to the consequent displacement of people due to environmental causes.

Human influence on the acceleration of climate change is widely recognized, albeit not in full agreement. This association is scientifically confirmed through the establishment of complex relationships between the level of GHG emissions, the frequency of extreme weather events and the intensity of the impacts generated. Evidence shows that the richest 10% of the world's population are responsible for half of global emissions (Ferreira, 2020, p. 40), while at the same time the poorest half produce only 10% of total emissions. At the same time, it is the poorest populations in the least emitting countries that feel the impacts of extreme weather events the most, which in turn are the result of human action characteristic of the Anthropocene.

According to the Global Risks Report 2024, World Economic Forum (2024), of the 34 long-term risk factors considered worldwide, 6 relate to the environment, with extreme weather events coming first, followed by critical changes in terrestrial systems (2nd), loss of biodiversity and the collapse of ecosystems (3rd), scarcity of natural resources (4th), pollution (10th) and non-climatic natural disasters (33rd).

Human activities are estimated to be the main contributor to the 1.0ºC increase in air temperature since 1987 (IPCC, 2019, p. 7), the year in which the *Our Common Future Report* was published, coordinated by Commissioner Gro Harlem Brundtland and produced by the World Commission on Environment and Development. This report formalized the concept of sustainable development from a systemic and global perspective, based on a set of values such as intergenerational solidarity. This issue is so important that in 2015, at the New York Summit (Ferreira, 2020, p. 44; World Economic Forum, 2024, p. 64), the 2030 Agenda and its Sustainable Development Goals (SDGs) were agreed, following on from the previous Millennium Development Goals (MDGs). In this context, the United Nations declared the 2020–30 to be the Decade of Action. The importance of environmental issues has been recognized worldwide, and the following stand out: SDG 13 – Climate action; SDG 6 – Drinking water and sanitation; SDG 14 – Protect marine life; and SDG 15 – Protect terrestrial life.

The global climate issue affects the scope of sustainability, posing threats when a new extreme event emerges. Valuing the planet, the physical space inhabited by humanity, and making it dependent on the climate highlights the urgency of action, a position that from an international point of view is also reinforced by the annual regularity of the debates framed by the United Nations Framework Convention on Climate Change (UNFCC) and the Conferences of the Parties on Climate (COP).

The IPCC (2022, p. 9) has shown evidence of human influence worsening the impacts produced by extreme weather events, which over time have been occurring with greater frequency and intensity. Table 5.1 shows the main estimated risks resulting from a rise in air temperature of 1.5ºC or 2.0ºC. It should be emphasized that in either situation the impact on populations and ways of life poses threats to survival.

Table 5.1. Climate Risks of a 1.5ºC or 2ºC Rise in Temperature.

| **Indicator** | **1,5ºC** | **2,0ºC** |
|---|---|---|
| Extreme events | 100% increase in flood risk | 170% increase in flood risk |
| Species | 6% of insects, 8% of plants and 4% of vertebrates affected | Affecting 18% of insects, 16% of plants and 8% of vertebrates |
| Urban water availability by 2100 | 350 million people affected by severe droughts | 410 million people affected by severe droughts |
| Population affected by heat waves every 20 years | 9% of the world's population (700 million people) | 28% of the world's population (2 billion people) |
| Population affected by sea level rise by 2100 | 46 million people affected by a 48 cm rise in sea level | 49 million people affected by the 56 cm rise in sea levels |

*Source:* Adapted from IPCC (2018).

Variations in the climate result in the confluence of more frequent and prolonged droughts with the loss of water resources, affectation of diverse species, and changes of the life models of local populations in human settlements fixed on the coastline, as a result of rising sea levels.

The combination of explanatory elements from the naturalistic and anthropogenic approaches is guided by an integrated and systemic view of the geographical and socioeconomic dimensions. It is understood that the world, as well as being tending to be unequal in its climatic, geomorphological, ecosystem, human, and social characteristics, produces GHGs in different ways and also feels their effects disproportionately. The countries of the North are, as mentioned above, the biggest emitters of GHGs, while the countries of the South make a minimal contribution to this problem, and in some cases their contribution is residual. However, these countries are the main and most directly affected by climate change, as have the least capacity to adapt. They are financially and economically poor countries, although they can demonstrate the availability and abundance of natural resources, but do not have the capacity to monetize them on a persistent basis.

Extreme weather events pose increased risks in well-identified and studied areas: poor countries; densely populated regions; Small Island States (SIDS); countries with long coastlines or riverside regions where there are concentrations of people. The regions most vulnerable to the impacts of extreme weather events coincide with experiences of uncertainty. It is common to consider that they simultaneously face the risk of a catastrophe, which results from the impossibility of predicting an event, and the inability to regulate the effects after it has occurred. The particularization of island territories, particularly micro-states – the much-studied SIDS – with long coastlines, oriented toward the ocean or with human settlements at the confluence of river basins. These are so-called fragile and vulnerable spaces where populations face uncertainty in their daily lives (Brito, 2024). Given the geographical location of these countries, which are naturally surrounded by oceans and have significant forest areas, despite the degree of impact they face, they have residual levels of emissions and are often defined as carbon sinks, helping to generate regional and global balances. An example of this is the case of the Democratic Republic of São Tomé and Príncipe, an African microstate located in the Gulf of Guinea, which is considered a carbon sink. It does not meet the carbon quotas, has low levels of emissions, but has already been particularly affected by rising sea, losing 4% of its physical territory.

From a global perspective, the degree of socio-environmental impact of extreme weather events is proportional to vulnerability, which represents a reduced capacity to adapt (CCPI, 2022; IPCC, 2022, pp. 9–12; Sui, 2022). From an environmental point of view, fragile ecosystems and endangered and endemic species will suffer the most from the losses and damage caused. For human communities, the impact is most evident in groups that are characteristically vulnerable and live in poverty, possibly extreme poverty or below the poverty line. These people lack protection and live in dependence on nature to ensure their daily survival and income.

Regional vulnerability is also differentiated. According to the Climate Change Vulnerability 2017 report (Verisk Maplecroft, 2016), of the 54 countries on the African continent, 41 have extreme or very high vulnerability and 12 moderate vulnerability. In Central America and the Caribbean, nine countries face extreme or very high vulnerability and 12 are at moderate risk. In South America, seven countries have very high vulnerability and three have moderate vulnerability. In Oceania, six live in extreme or very high vulnerability and two in moderate vulnerability. On the Asian continent, 14 face extreme or very high vulnerability and 32 moderate. Only North America and Europe present a low risk of vulnerability, as these are the regions with the greatest capacity to adapt. If the analysis focuses on the prevailing climate patterns, it is easy to see that the highest risk areas coincide with the countries located in the intertropical zone, influenced by the Equator. These are the regions most affected by three types of extreme weather events: the characteristically tropical storms marked by high temperatures and intense rainfall, floods and hydrological phenomena, but also droughts associated with extreme temperatures and forest fires.

In the global context, the existence of climate vulnerability is now unquestionable, resulting from the combination of exposure and sensitivity, reducing the capacity to adapt (Fig. 5.1). Climate change creates risks and impacts on ecosystems and their resources, as well as on human communities. But on the other hand, latent or emerging extreme weather events are also aggravated by human activity through GHG emissions, making this a circular problem.

As times have changed, awareness of the urgency of climate change and the urgency of action have taken shape, requiring the adoption and implementation

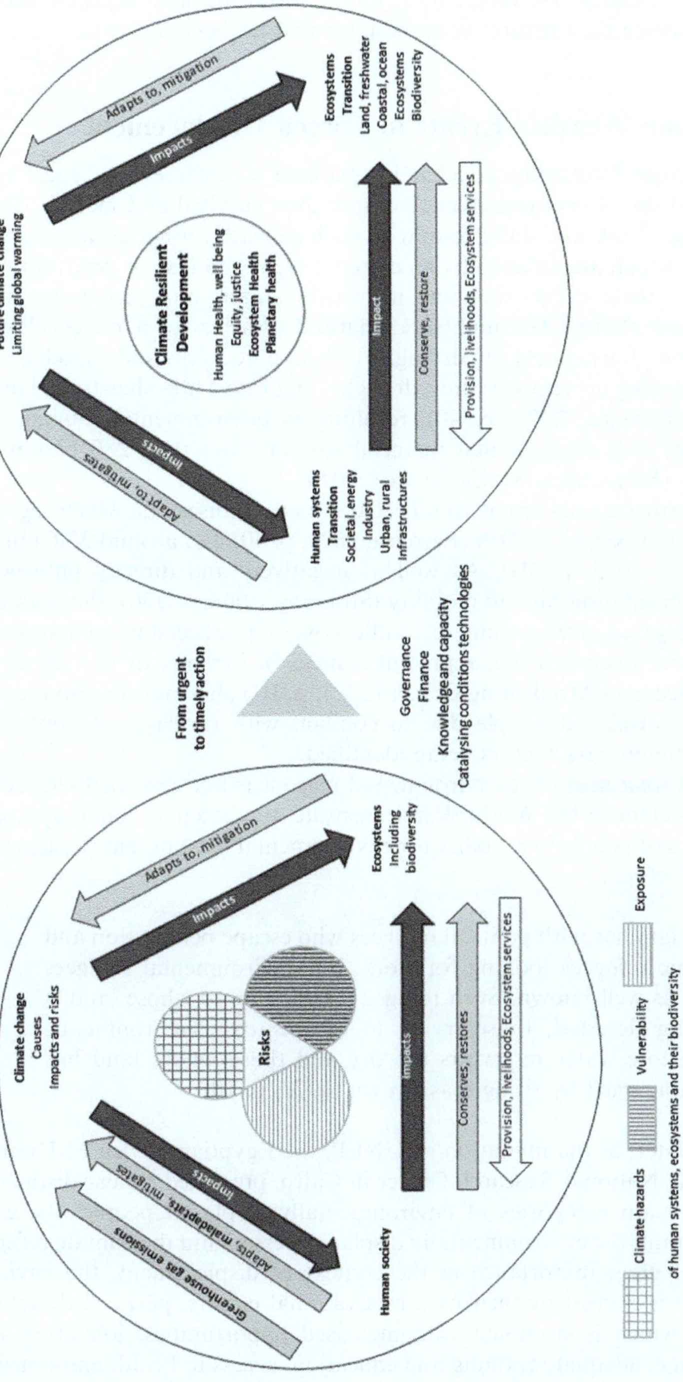


Fig. 5.1. From Climate Risk to Resilient Development. *Source:* Created by the author from the original (IPCC, 2022).

of adaptation measures. However, mitigation actions are also urgently needed worldwide, favoring the creation of sustainable and resilient societies.

## From Extreme Weather Events to Forced Displacement

In most territories frequently affected by extreme weather events, local populations depend on natural resources to ensure their survival and income. These are mostly basic, vital, and daily resources, such as water, fisheries, and agricultural products, which are affected by an extreme event and lose or deteriorate. In many countries, these events represent irreversible conditions, generating situations of unsustainability. The number of natural disasters of a meteorological (tropical storms, hurricanes), hydrological (heavy rainfall and floods), and climatological (rising air temperatures, droughts and forest fires) has tripled in the last 30 years (Ferreira, 2020, p. 40), resulting in environmental, human and property losses, with an estimated financial loss of more than 280 billion US dollars in 2021 (MunichRE NatCatService, 2020).

Extreme weather events are often reflected in natural disasters, which together affected 40 million people in 2019 (Ferreira, 2020, p. 40) and around 32.6 million in 2022 (IDMC, 2023, p. III). As well as negatively and directly influencing livelihoods by promoting human mobility (Stojanov, 2008, p. 130), they can also lead to the emergence or worsening of conflicts over the ownership, management, and use of scarce resources that represent sources of income. In this sense, the Internal Displacement Monitoring Center (2023, p. III) also indicates an increase of 28.3 million displaced people due to conflict, with no cases of confluence between the different risk factors being identified.

Analyzing displacement for environmental reasons is not new. In 1976, Lester Brown, then director of the World Watch Institute, presented an initial approach to the categories of people in mobility for environmental reasons, emphasizing the environmental refugee:

> We are familiar with political refugees who escape persecution and economic refugees looking for work, but environmental refugees are not as well known. Such refugees include those whose land is becoming deserted, those trying to escape toxic environments, those whose water reservoirs are dry and those whose land has been submerged by rising seas Brown (1977, p. 10).

Nine years later, at the invitation of UNEP, the Egyptian Essam El-Hinnawi (1985), from the National Research Center in Cairo, presented a clear distinction between three main categories of environmentally displaced people: the environmental migrant; the environmentally displaced person and the climate refugee.

Although of great importance in the context of displacement, the environmental migrant is framed by their own motivational factors, particularly related to well-being, which is connoted with increased opportunities, job offers and improved income, adequate training and education, access to health and housing.

Environmental migration, just like any displacement within the framework of the migratory process, is promoted by choice and will and is initially defined in a conventional way by economic criteria.

As far as the environmentally displaced person and climate refugee are concerned, nature is an added burden which, by becoming hostile, makes it impossible for them to remain in their place of origin since it is not compatible with maintaining life. El-Hinnawi (1985, pp. 23–28) identifies environmental factors, of both natural and anthropogenic origin, which result in a socio-environmental process of repulsion, pushing and forcing populations to move, whether internally or across borders. The progressive and continuous transformation of ecosystems, such as deforestation promoted by the timber industry for the international market, in sensitive forest contexts alters the ecosystem pattern and promotes climate changes so profound that often become irreversible. This is the case of tropical forests, such as the Brazilian Amazon, the Gulf of Guinea forest park or the forests of south-east Asia, which were originally classified as primary and have become secondary as a result of massive deforestation. In these cases, the aim is the appropriation and sale of hardwoods or the introduction of fast-growing substitute crops such as soya, both oriented for the international market. Either process results in a loss of the soil's productive capacity which, in the long term, promotes desertification, drought and consequently food insecurity and hunger. The category of environmental refugees can be defined as:

> those people who have been forced to leave their traditional habitat, temporarily or permanently, because of a marked Environmental disruption (natural and/or triggered by people) that jeopardized their existence and/or seriously affected the quality of their life El-Hinnawi (1985, p. 4).

Environmental displacement can be analyzed on the basis of a set of five factors outlined by Gregory White (2011), which are based on the influence of extreme climatic events affecting people's lifestyles at a local level. Through its manifestations, climate change:

(1) Influence and affect the availability of vital resources, conditioning access, and utilization capacity. This is the case, for example, of river basins which, in the case of an extreme climatic event, either due to excess or failure, suffer from a change in flow. In these cases, the water can become contaminated with dust, sand and various bacteria, making it impossible to consume it once has been collected.

(2) Constrain production rhythms and limit harvests, resulting in a loss of food security and increasing hunger situations. Regardless of whether the extreme weather events are meteorological, climatological or hydrological, the reduction in water availability or the devastation of cultivated productive fields has repercussions on agricultural losses with implications for the food patterns of local populations.

(3) Influence local and national management models, as well as the control of natural resources, a situation that is more marked when spaces and resources are shared between communities or ethnic groups of different origins. This situation implies a capacity for negotiation which, at critical moments, is not easily demonstrated.
(4) Promote a change in economic performance at national and regional level, characterized by variations in demand for goods and services in the face of unavailable supply. In contexts marked by poverty and vulnerability, these events indicate the emergence of aggravated fragilities that can precede phases of economic recession and crisis.
(5) Facilitate the emergence, and worsening, of situations of community and ethnic conflict whenever there is a dispute over scarce and vital resources, or the control of sources of income.

Lester Brown (2011, pp. 77–82) resumed his analysis of the issue of climate refugees by presenting a classification based on the most relevant types of extreme events that affect the greatest number of people:

- refugees from rising sea levels, particularly small island territories such as Tuvalu, Kiribati, the Marshall Islands and the Maldives. In the case of SIDS that emit or potentially promote climate refugees, the case of Tuvalu is the most paradigmatic, where it is estimated that 3,000 inhabitants have started displacing to New Zealand as a result of rising sea levels (Idem, 2011, p. 74);
- refugees from global warming, since it has been scientifically proven that higher sea surface temperatures encourage frequent and intense extreme meteorological events, particularly tropical storms. In this case, the most common examples are Central America and the Caribbean, South-East Asia and the Gulf of Bengal (Idem, 2011, p. 76);
- refugees from desertification and the advance of deserts, which progressively make sedentary human life precarious and even unviable. The most common cases are dry territories, such as the dry corridor in Central America, or those located in areas bordering deserts, such as the Sahara, which has spread in all directions, affecting neighboring countries (Idem, 2011, p. 77);
- refugees from water scarcity, which result from the decreasing flow of river basins and water tables, and the possible depletion of aquifers, promoting social situations of hydrological poverty. The most typical cases of this are north-west India, northern China and Mexico (Idem, 2011, p. 78);
- refugees from toxic waste and radiation levels, which occur mainly in areas affected by toxic chemical waste that has not been properly treated, or nuclear accidents with the spread of radioactivity. The most common example, due to the degree of destruction it caused, was Chernobyl, which affected more than 350,000 displaced people (Idem, 2011, p. 80).

As Essam El-Hinnawi (1985, pp. 28–30) points out, corroborated by Gregory White (2011) and Will Steffen et al. (2015), there is a circular and systemic interpretation of the socio-environmental relationship that triggers human displacement. Although Paul Crutzen and Stoemer (2000) analyzed the problem from a purely causal perspective, the concept of the Anthropocene as a human era that negatively influences the environmental framework and climatic conditions is recovered. It is humanity that interferes with the environment in an intrusive and unplanned way, without foreseeing the possible impacts, degrading and depredating. Likewise, and following systemic thinking, it is humanity that feels the effects of climate variations most directly and with the greatest impact.

The knowledge of the existence of a large population worldwide that is faced with the need for protection due to exogenous factors such as climate, to the point of being conceptually defined as displaced persons or climate refugees, highlights the inconsistency of the legal framework of international law. The treatment given to people in these circumstances does not go beyond the dimension of migration, since refugee status is defined under the Magna Carta of the Geneva Convention on the Status of Refugees, which dates back to 1951, but does not include environmental factors that pose a risk or threat to life.

From the perspective analyzed of people who, regardless of the regional area, suffer from the intensity and durability of the impacts of extreme weather events, preventing them from remaining in their places of origin and forcing them to move around marked by uncertainty, the idea of escape and the need for refuge seem obvious. Even so, and despite multilateral awareness of these increasingly frequent and intense processes, the revision of the Magna Carta on Refugees is still not a reality, leaving thousands of people without international protection.

## Systematizing Without Concluding

In the context of the 21st century, talking about the risks and challenges of climate change is an exercise to identify causes, define impacts and, above all or primarily, design models to minimize the signs of an emerging problem.

An evolutionary reading, framed by global initiatives, helps us to understand that the causes of the climate crisis and the social and human effects are easily identified and categorized. The impacts are evident today, both from an environmental point of view due to the inherent destruction, in economic terms due to the losses caused by recessions and imminent crises, but also in terms of the social dimension reflected in the uncertainty of life models and, very importantly, in mass displacement.

Independently of the regional area that is analyzed, it can be considered that it is predominantly anthropogenic factors that worsen environmental and natural conditions by modifying the patterns of ecosystems that sustain human life, promoting different forms of displacement. It is in this context that environmental migrants, displaced persons and climate refugees come to the center: due to the lack of resources such as fresh water, arable land and pasture available to guarantee survival. These are also factors that encourage the emergence of conflicts associated with the ownership, management and use of scarce resources which, as well as guaranteeing survival and family maintenance, represent the most important sources of income. These factors emphasize fragility and, at the same time, the need for protection, whether national or international.

As far as resolution models are concerned, doubts persist, but the solutions need a global approach based on the assumption of shared and proportional responsibility. It is within the framework of accountability that the main challenges are defined without being resolved.

The attempts to find a common point of intervention between all stakeholders are conceptually and theoretically defined, but there is no effective operationalization to guarantee a slowdown in extreme weather events, control of the effects and the settlement of populations in their places of origin. The most vulnerable populations continue to feel the effects and continue to displace, either voluntarily through migration or forced through internal and external displacement, albeit without any protection.

Finally, although it is not entirely possible to conclude, it can be remarked that climate change and its extreme events pose multiple risks to all sectors of life. The resulting impacts are understood as challenges that remain unsolvable, and the same applies to the situation of people fleeing and asking for protection, whether national or international, without any response due to the inconsistency of the status that has been in force, and assumed, since 1951 without any adjustment to the evolution of the times.

## References

Brown, L. (2011). World on the edge. *How to prevent environmental and economic collapse*. Earth Policy Institute. W. W. Norton & Company.

Brito, B. (2024). A fragilidade dos Estados em África: uma condição que persiste. In *Janus2023, Anuário de Relações Exteriores*. Universidade Autónoma de Lisboa.

Brown, L. (1977). *Redefining national security*, Worldwatch Paper n. 14. Worldwatch Institute. https://files.eric.ed.gov/fulltext/ED147229.pdf. Accessed on January 20, 2024.

CCPI. (2022). *Climate change performance Index*. Germanwatch. https://ccpi.org/wp-content/uploads/CCPI-2023-Results-3.pdf. Accessed on November 10, 2023.

Crutzen, P. (2002). Geology of mankind. *Nature*, *415*, 23. https://www.nature.com/articles/415023a.pdf. Accessed on January 15, 2024.

Crutzen, P., & Stoemer, E. F. (2000). International Geosphere Biosphere Programme (IGBP). *Newsletter*.

El-Hinnawi, E. (1985). *Environmental refugees*. United Nations Environmental Programme.

Ferreira, P. M. (2020). *Desenvolvimento e Alterações Climáticas. Impactos e (in)sustentabilidade*. Policy Paper, Dezembro. FEC.

Fundação Oceano Azul. (2021). *A emergência climática e a extinção de espécies*. Factos e números. Fundação Oceano Azul.

IDMC. (2023). *Internal displacement and food security*. Internal Displacement Monitoring Centre. Norwegian Refugee Council.

IPCC. (2018). *Special report. Global warming of 1.5ºC*. Intergovernmental Panel on Climate Change. https://www.ipcc.ch/sr15/. Accessed on January 20, 2024.

IPCC. (2019). *Aquecimento Global de 1,5°C.* Relatório especial do Painel Intergovernamental sobre Mudanças Climáticas. PNUD, GEF, Ministério da Ciência, Tecnologia, Inovações e Comunicações do Brasil.

IPCC. (2022). *Climate change 2022: Impacts, adaptation and vulnerability.* Contribution of Working Group II to the sixth assessment report of the Intergovernmental Panel on Climate Change. Cambridge University Press. https://doi.org/10.1017/9781009325844

Maldonado, M. A. (2018). *Antropoceno: La política en la era humana.* Barcelona.

Maplecroft. (2017). *Climate change vulnerability index 2016.* Verisk Maplecroft.

McNeill, J. R. (2000). *Something new under the sun: An environmental history of the twentieth-century world.* W.W. Norton & Company.

Morin, E. (2007). *Introdução ao pensamento complexo.* Sulina.

MunichRE NatCatService. (2020). https://www.munichre.com/en/solutions/for-industry-clients/natcatservice.html. Accessed on October 25, 2023.

Pultarova, T. (2022). Milankovitch cycles: What are they and how do they affect Earth? *Space.com.* https://www.space.com/milankovitch-cycles. Accessed on January 10, 2024.

Santos, F. D. (2021). *Alterações climáticas.* Fundação Francisco Manuel dos Santos.

Steffen, W., Broadgate, W., Deutsch, L., Gaffney, O., & Ludwig, C. (2015). The trajectory of the Anthropocene: The great acceleration. *The Anthropocene Review, 2*(1), 81–98.

Stojanov, R. (2008). *Development, environment and migration: Analysis of linkages and consequences.* Palacký University.

Sui, C. (2022). The "problem" with the climate change problem. https://insight.factset.com/the-problem-with-the-climate-change-problem. Accessed on November 3, 2023.

Verisk Maplecroft. (2016). Climate change vulnerability 2017. https://www.maplecroft.com/global-risk-data/climate-risk-dataset/. Accessed on October 25, 2023.

White, G. (2011). *Climate change and migration. Security and borders in a warming world.* Oxford University Press Inc.

World Economic Forum. (2024). The global risks report 2024. https://www.weforum.org/publications/global-risks-report-2024/. Accessed on February 20, 2024.

World Meteorological Organization. (2023). *State of the global climate 2022.* WMO No.1316. World Meteorological Organization.

World Resources Institute. (2021). Global greenhouse gas emissions 2016 excluding land-use change and forestry (LUCF) from Climate Watch. https://www.climatewatchdata.org/ghg-emissions?chartType=percentage&sectors=614. Accessed on November 18, 2023.

Chapter 6

# Need for a Human Rights-Based Approach to Climate Change Adaptation and Mitigation With Reference to the State of Uttarakhand in India

*Amit Upadhyay and Abhinav Mehrotra*

O.P. Jindal Global University, India

## Abstract

This chapter analyzes whether a human rights approach can serve as a model for addressing climate change issues in India. It discusses the debate between the compatibility of aims in addressing climate change and realizing human rights, focusing on climate change issues in Uttarakhand in India. It further discusses the absence of sound adaptation policies, pointing to the violations of the social and economic rights of the populations residing in the affected regions and suggests a tangible legal framework for analyzing state action on adaptation policy in response to climate change by referring to the successful case study of early warning system in the Netherlands. The authors conclude that any strategy to deal with climate change concerning adaptation or mitigation must foresee the consequences for humans, as part of communities and individually, for which the human rights framework assumes significance.

*Keywords*: Climate change; adaptation; mitigation; human rights; Uttarakhand and India

## The Human Rights-Based Approach

The United Nations was created in 1945 with three primary purposes: maintaining international peace and security, economic development, and protecting and promoting human rights (Baehr, 2004). An essential part of protecting and

**The Social Consequences of Climate Change, 67–78**

**doi:10.1108/978-1-83797-677-520241006**

promoting human rights includes protection against any loss and damage from climate change. Given the significance of climate change, it has been labeled as the human rights challenge of the 21st century (Touissant & Blanco, 2019). The discussion surrounding the loss and damage caused by climate change has been gaining prominence due to its impact on a wide range of internationally recognized human rights. However, the relationship between loss and damage and human rights remains significantly understudied (Schapper, 2020). Loss and damage may be defined as the "negative effects of climate variability and climate change that people have not been able to cope with or adapt to" (Humphreys, 2009). As a result, the impact of loss and damage violates human rights in different ways, including violation of civil and political rights, such as the right to life, liberty and property, as well as economic, social, and cultural rights, such as the right to work, education, social security, the highest attainable standard of physical and mental health, adequate food, clothing and housing, and collective rights, including the right to development, self-determination, peace, a healthy environment, and minority rights which necessitates the adoption of human right law-based approach to address the human rights implications of loss and damage resulting from climate change (Daniel, 2010).

The human rights law-based approach to climate change builds on the existing obligations of parties under international and regional human rights treaties that they have ratified. It provides a method for integrating human rights within climate change policies that are not merely confined to certain kinds of human rights. This approach considers the integration of loss and damage in the work of relevant human rights bodies, including the United Nations Human Rights Council and the Office of the High Commissioner for Human Rights, to integrate the normative dimension of human rights into the design, implementation, and evaluation of policies and actions concerning climate change (Sinden, 2007).

The response under international law to climate change through the United Nations Framework Convention on Climate Change (UNFCCC), which arguably has not been able to integrate human rights into policy outcomes, has resulted in a lack of protection for communities affected by loss and damage due to climate change (Savaresi & Auz, 2019).

In this light, this chapter proposes adopting a human rights-based approach to address the issue of climate change to strengthen the response to loss and damage at the international and national levels.

## International Law Dealing With Climate Change and Human Rights

Under international law, the issue of climate change has been dealt with under various international conventions, including the United Nations Framework Convention on Climate Change and the Paris Agreement. As per the preamble of the United Nations Framework Convention on Climate Change, climate change is a "common concern for humankind" that "may adversely affect natural ecosystems and humankind" (UNFCCC, 1992; preamble).

Under the UNFCCC, the concept of human rights found a special mention when the G-77 recognized the right to development in Article 2 as an "inalienable human right," which then was resisted by the industrialized countries (Duyck et al., 2018, p. 4) and could not find a place under the covenant. The reason behind the resistance to respect human rights was to see both the fields of human rights and climate change as distinct.

However, with the advent of the Paris Agreement, human rights issues gained prominence under climate governance by referring to specific rights, including the right to development. Its preamble states the following:

> Parties should [. . .] respect, promote and consider their respective obligations on human rights, the right to health, the rights of indigenous peoples, local communities, migrants, children, persons with disabilities and people in vulnerable situations and the right to development, as well as gender equality, empowerment of women and intergenerational equity. Paris Agreement (2015), preamble

Even before the Paris Agreement in 2003, multiple bodies under the UN had come up with the framework the authors in this chapter are arguing for, i.e., the Human Rights approach. The aim in 2003 was to integrate human rights within the working of UN agencies working toward realizing the right to development (UNGA, 2006). Specifically, these UN bodies outlined six guiding principles which should be the basis upon which the development policies must adhere, including universality and inalienability, indivisibility and interdependence and interrelatedness of human rights; equality and nondiscrimination; participation and inclusion; and accountability and the rule of law.

The necessity for taking human rights into account in the implementation of mitigation and adaptation measures has even been recognized by the Conference of Parties to the UNFCCC, which has stated that, in all climate-change-related activities, states parties should fully respect human rights (UNFCCC, 2010).

In this context, the Human Rights Approach may be defined as a "conceptual framework for the process of human development that is normatively based on international human rights standards and operationally directed to promoting and protecting human rights," particularly about climate mitigation and adaptation (Duyck et al., 2018, p. 4).

Through this approach, it is possible to ensure that laws, policies, programs, and projects that have been adopted to mitigate or adapt to climate change respect, protect, and fulfill human rights under international human rights law to provide judicial recourse for loss and damage caused by the climate change. Consequently, it empowers the affected communities and their representatives through the civil societies to put forth public opinion and grievances.

Despite having in place the provisions under the instruments mentioned above, the implementation of the aims of promoting and protecting human rights violated due to climate change has not taken place at the municipal level, especially regarding mitigation and adaptation due to various factors, including the national priorities, democratic conditions, the level of enforcement of relevant national and international human rights treaties that have been ratified (Lesnikowski et al., 2023). This necessitates a further analysis of climate change as a human rights issue in the next section.

## Climate Change as a Human Rights Issue

The Intergovernmental Panel on Climate Change (IPCC) was created in 1988, and the subsequent creation of the UNFCCC in 1992 coincided with the collapse of the Soviet Union, which brought significant changes in human rights, leading to the Vienna Declaration of 1993. The Vienna Declaration did emphasize that national and regional differences must be considered, but this did not exempt the states from their duty to promote and protect all human rights and fundamental freedoms (Vienna Declaration and Programme of Action, 1993). At the same time, the Asian Regional Meeting ended by issuing the Bangkok Declaration, which defined and summarized the idea of cultural exception to human rights in terms of considering the specificities and historical, cultural, and religious backgrounds of various regions (Bangkok Declaration, 1993).

By applying this declaration in the context of climate change, adaptation is necessary for all the states, even if the specific social, cultural, and religious backgrounds are considered to address the impacts of climate change, as adaptation is the only available and appropriate response. Along with mitigation measures to achieve sustainable development and improve the economic condition of the most vulnerable to climate change, the poor.

Over time, several states have come to recognize that climate change is another international crisis that has profound implications on human rights that find place under various international covenants, declarations, treaties, and agreements, especially concerning environmental and human rights that include the Universal Declaration of Human Rights (UN, 1948); International Covenant on Civil and Political Rights (Office of the High Commission for Human Rights, 1966a); International Covenant on Economic, Social and Cultural Rights (UN, 1966b); Convention on the Rights of the Child (UN, 1989); Declaration on the Rights of Indigenous Peoples (UN, 2008). To illustrate, under Article 3 of the Universal Declaration of Human Rights (UN, 1948): "Everyone has the right to life, liberty and security of person." The intrinsic significance of life, liberty, and security of a person is recognized increasingly coupled with the quality of the environments in which people live, which will be reflected upon in the next section in the context of the State of Uttarakhand, especially concerning the issues of mitigation and adaptation.

Under the obligation to protect human rights, states have an international and domestic duty to intervene to prevent acts of third parties threatening human rights and punish actual violations. As per Article 11(1), ICESCR states have to take the necessary measures to ensure adequate living conditions (UN, 1966b). Similarly, under Article 12 ICESCR, which links health and environment and bestows responsibility upon the states parties to recognize the right of everyone to the enjoyment of the highest attainable standard of physical and mental health and to take steps to achieve the full realization of this right shall include the improvement of all aspects of environmental and industrial hygiene (UN, 1966b).

Historically, a group of small island developing states had signed the Malé Declaration on the Human Dimension of Global Climate Change (UNHRC, 2009) in the Maldives. Through the declaration, the UNFCCC was requested to collaborate with the UNHRC to study the impact of climate change on human rights in general and the right to health in particular. This declaration is highly significant to prioritize climate change as a human rights issue as it laid down a roadmap for actions within the UN system designed to explore and draw attention to the relationship between global warming and the full enjoyment of human rights, both civil and political, as well as economic, social, and cultural rights. The questions that this declaration explored include the complex and multifaceted relationship between climate change and human rights, whether climate change constitutes a violation of human rights, especially the rights of vulnerable people, and what are states' national-level and international-level human rights obligations pertaining to climate change.

A significant development took place on April 19, 2023, when the General Assembly requested an advisory opinion on the obligation of states to respect climate change based on a resolution (ICJ, 2023) that was adopted on March 29, 2023, in the UNGA. The issues brought before the International Court of Justice by the UNGA included the obligations of states under international law to ensure the protection of the climate system and other parts of the environment from anthropogenic emissions of greenhouse gases for states and for present and future generations.

At this stage, it needs to be emphasized that the obligations to respect, protect, and fulfill human rights compel states to take mitigation and adaptation measures against climate change. Concerning nonstate, natural, and legal persons, a more correct human rights framework to approach State mitigation measures through the obligation to protect human rights for states implies a duty to prevent and punish human rights breaches caused by third parties and is a positive obligation upon them.

At the same time, the obligation to fulfill human rights, which is positive and compels states to facilitate, provide, and promote the enjoyment of fundamental claims, is also relevant to mitigation measures to the extent that it can be interpreted as a means of facilitating the realization of fundamental rights which is particularly important for adaptation measures that aim to react to the impacts of climate change by implementing national strategies (Duyck et al., 2018, p. 4).

Overall, with regard to both mitigation and adaptation measures, states have an obligation to respect the principle of nondiscrimination, which underpins all human rights, particularly based on the right to equal protection before the law.

All the obligations discussed above are relevant in establishing mitigation measures within the context of the evolving human right to development, entangled with several other human rights, such as food, water, and health. However, the implementation of the same in developing countries like India is challenging given the low level of development that raises obstacles to fight climate change.

## Human Rights and Climate Change in Uttarakhand, India

India has been a Party to the United Nations Framework Convention on Climate Change, its Kyoto Protocol, and the Paris Agreement. Over the years, it has taken several steps to overcome the hurdle of climate change as a party to the said conventions. In 2008, India released the National Action Plan on Climate Change (NAPCC) that outlined India's strategy to meet the challenge of climate change. The focus of the NAPCC was on adopting mitigation measures through the use of clean and green technologies for development purposes, with an emphasis on maintaining a high growth rate (Pandve, 2009). India, through its Nationally Determined Contribution (NDC), has aimed to balance the concerns and priorities of climate change, sustainable development, including poverty eradication, and economic growth of the country. In August 2022, India updated its NDC, according to which India has an enhanced target to reduce the emissions intensity of its GDP by 45% by 2030 from the 2005 level (NAPCC, 2008).

Adaptation is particularly important in India, which is the case in most developing countries. Adaptive capacity may be defined as the ability of a system to adjust to climate change (including climate variability and extremes), to moderate potential damages, to take advantage of opportunities, or to cope with consequences (Adger et al., 2003).

Coming to the issues faced by a developing country like India due to climate change and its intersection with human rights, the State of Uttarakhand has seen several natural disasters caused by climate change, such as floods. For example, the Chamoli district in the state of Uttarakhand in India suffered from floods (Verma et al., 2022). The disaster occurred in the upper catchment of the *Rishiganga* River, a tributary of the *Alaknanda*, which led to a sudden rise in the water level of the Rishiganga that destroyed the *Rishiganga* hydropower project as well as the *Tapovan Vishnugad* hydroelectric project.

Although, the judiciary in this state has been proactive in protecting the rights violated by climate change. In March 2017, the High Court of Uttarakhand in India ruled in the Ganges and *Yamuna* rivers cases that the Gangotri and *Yamunotri* glaciers, along with these rivers, as well as other natural objects in the state of Uttarakhand, enjoy legal rights (Donnell, 2018). The High Court established the natural objects as legal minors and conferred guardianship responsibilities on several individuals within the state government of Uttarakhand.

In the first judgment, concerning the Ganges and Yamuna case (High Court of Uttarakhand, 2017), the High Court declared the two rivers and all their tributaries' juristic/legal persons/living entities having the status of legal persons. In this case, the Court declared a group of officials, including the Chief Secretary of the State of Uttarakhand and the Advocate General of the State of Uttarakhand persons in *loco parentis* as the human face to protect, conserve and preserve Rivers Ganga and Yamuna and their tributaries.

Meanwhile, for the second case that relates to the Glaciers, a petition was initiated by an advocate of the same High Court. The petition argued for extending legal personhood to all other natural objects in the State of Uttarakhand, including the *Gangotri* and *Yamunotri* glaciers. Similarly, like the previous case, the Court ordered the Chief Secretary of the State of Uttarakhand, together with legal advisors, academicians, and judges, to act as the persons in *loco parentis* as the human face to protect, conserve and preserve all the Glaciers (High Court of Uttarakhand, 2017).

Both cases were significant as they created legal rights for natural objects in Uttarakhand. Such decisions have also been given in jurisdictions like New Zealand, which recognized the Whanganui River and its catchment as a legal person (Whanganui River Claims Settlement Act, 2017). The difference between New Zealand and India stems from the fact that in New Zealand, the legal rights were created through legislation, whereas the two cases discussed in India were judicial responses to the public interest litigation to protect natural objects.

Coming back to India, although the Indian Constitution recognizes an obligation of the state to endeavor to protect the environment, including the rivers, glaciers, and other natural objects. Through these cases, the ability to take legal action provides an important tool for the protection of the environment in Uttarakhand. Nonetheless, the challenges pertaining to the implementation and enforcement of legal rights for nature continue to take place, requiring steps to mitigate and adapt to situations where climate change-induced occurrences take place. The reason is that the impact of climate change in India varies from region to region as some of them experience rainfall, flood, and storm risks while other regions encounter sparser rainfall and prolonged droughts, the state of Uttarakhand falling in the former category (Sathaye et al., 2006).

As Uttarakhand is a mountainous state, especially for remote communities, they are more vulnerable to climate change impacts than nonmountainous areas. There are several challenges that are faced in combating climate change, including a lack of adequate data in terms of quantity and quality about climate change impact, weak institutional capacity at various governance levels, social and economic barriers, and poor infrastructure for adaptation purposes in such areas. Additionally, these areas are short on lifesaving and livelihood-supporting infrastructure and lack access to climate information and support services. On the other hand, despite all the aforementioned limitations and obstacles, one key advantage in these areas is the community support that has helped them survive the vicissitudes of nature. Through these community structures, a human rights-based approach can bring together the local knowledge systems and management practices that can usefully inform adaptation responses (Agarwal, 2008).

Despite the significance of local knowledge, the documentation of the same is limited and, as a result, has not been scientifically proven. This mishandling is largely attributed to the institutional hindrances faced by the nonstate actors in these regions whose rights are violated.

Thus, although climate change is the center of the issues faced by people, there are accompanying factors, including population change, economic growth, urbanization, and globalization, that have caused rapid socioeconomic differences. The current level of understanding of adaptation needs urgent interventions specific to mountainous regions, including, among other things, changes in institutions and institutional processes for better governance. The transformation of local institutions for human capacity building and for setting up technologically enabled systems and stakeholder partnerships involving communities and government agencies will play a crucial role in effective adaptation outcomes.

## Learning From the Early Warning System in the Netherlands

Going further, the Netherlands is another country that has faced risks caused by floods due to its geographical position. What is to be noted is the fact that the Netherlands has been successful in avoiding being a victim of continuous flooding and has recovered and prevented such occurrences by focusing on recovery from flooding and land reclamation. Although issues surrounding water safety have been part of their lives, the people of the Netherlands have developed embankments for centuries to increase flood prevention. Interestingly, to ensure regional cooperation, the Netherlands has The Safety Regions Act in place that seeks to achieve an efficient delivery of fire services, medical assistance, and crisis management under one regional management board. The Act aims to structure vulnerable regions based on the level of threat, as is the case with regions marked by police (Van Der Steen et al., 2018).

Their current water safety approach or flood risk management focuses on flood preparedness as well as flood prevention and mitigation, which is missing in the case of India, as explained through the case study of Uttarakhand. Disaster management in the Netherlands does not merely include efforts to develop an adequate disaster management system, but also to raise awareness and resilience, which is absent in India as the role of local knowledge has not been utilized. By using the tools of awareness and resilience, the Dutch have formed the Flood Management Taskforce, whose emphasis is on empowering society to handle the crisis by mitigating the impact of extreme weather events that will lead to the protection of such adversities.

Further, the Dutch government has put in place an early warning and alert system for disasters and life-threatening situations, which prepares people to take steps to face impending disasters. This practice has been proven to be beneficial in facing the challenges posed by climate change. The early warning system in the Netherlands has been highly successful and has saved the lives of the people, which, if established in India, may prove to be an integral component to combat climate change.

## Need for Adaptation Measures to Combat Climate Change in Uttarakhand, India

At the international level, the focus on the adaptation and mitigation of environmental and climate change has been put in place by adopting the Sustainable Development Goals and ratifying the 2015 Paris Agreement, to which India is also a party (UNFCCC, 2015). However, based on the analysis, in India, the challenges are varied and multifaceted and require consideration of the governance challenges that exist. As a result, the process of effective implementation is to be based on national-level strategies and steps.

To develop such policies, certain factors must be looked into through a human rights-based approach. The foremost requirement is to reduce poverty and improve the economic condition of the vulnerable regions by ensuring their economic and social rights, as a nation's adaptive capacity depends on the country's economic and social well-being. Further, the extent to which the availability and accessibility of technology exist leads to multiple options for adaptation measures such as warning systems, protective structures, relocation of human settlements, and flood control measures, as discussed in the context of the Netherlands. Implementing both these measures largely depends on the level of knowledge and access to information available to the citizens. By increasing knowledge and facilitating access to information through a human rights-based approach, the preparedness of the community and the nation increases, thus averting impending disasters. Thus, choosing the best suited adaptation method involves recognizing the need to adapt and identifying the available options. It is necessary to provide access to information not only to decision-makers but also to vulnerable communities. To ensure the same, the role of institutions assumes significance (Ojha et al., 2019).

As analyzed in the preceding section, the reason for the success in the Netherlands was the effective role of institutions in ensuring preparedness and timely action by developing adaptation strategies and disseminating information on climate change (Van Der Steen et al., 2018) which is missing in India, coupled with the lack of involvement in the decision-making processes of the citizens and their voice on protecting their human rights. This requires mainstreaming the environment to sensitize governance, institutions, and political systems to improve the environment that benefits the poor and reduce their vulnerability to the effects of climate change.

Finally, another significant indicator for analyzing the adaptive capacity is the equity in the allocation of resources and access to information across age, gender, and ethnicity to the marginalized communities and sections of a population that are vulnerable because of their inability to access information or resources. Thus, it is important to ensure that all sections of the population are equitably adapted to the effects of climate change.

## Steps to Be Taken Regarding Adaptation Measures to Combat Climate Change-Induced Disasters Nationally and Internationally

Globally, the rules and guidelines that address the response to climate change deal with the need to manage and respond to disasters, minimize the impact of disasters, and reduce disaster risks. Although, there is no comprehensive legal framework that guides international disaster law. There are regional laws and bilateral agreements that exist. This gap highlights the need to have a legal framework in place and enhance disaster law and its response both at the international and national levels, which requires putting in place mechanisms to ensure climate justice and globalize the right to protection from such disasters. To overcome such disasters, there is a need for shared action, cooperation, engagement, and partnership. Thus, there is a need to understand disaster risk, strengthen disaster risk reduction for resilience, invest in disaster risk reduction for resilience, and enhance preparedness for response (Ojha et al., 2019).

For India, what this means is that the mechanisms to collect relevant scientific data about the disaster, foresee and predict its occurrence, and ultimately disseminate the information to those at risk require a structure that protects the human rights of vulnerable people. Considering the need for awareness of the human rights dimensions of climate change in the context of India and internationally, the responsibility of states toward adaptation and mitigation for climate change should be reformulated. The reason is that the current treaties dealing with climate change do not create a direct obligation between states and nonstate entities. This question is particularly relevant for climate change adaptation measures. Furthermore, directly applying the obligations to respect, protect, and fulfill human rights would be more binding on states and would facilitate international cooperation (Limon, 2009).

Another perspective to investigate is to develop a potential human right to a Sustainable Environment within the category of third-generation human rights. The rationale behind it is the fact that human rights are vulnerable to environmental degradation and depend on a supportive environment (Wewerinke, 2013, pp. 9–12).

Thus, by combining both approaches, it seems possible that a correlative obligation to respect, protect and fulfill a sustainable environment would make it possible to hold a State responsible, which would directly trigger human rights protection mechanisms, particularly UN universal procedures.

Furthermore, such obligations would constrain the margin of discretion states enjoy in deciding climate change mitigation and adaptation policies. For some states, the regional human rights protection frameworks, like the European, Inter-American, and African systems, exist, but for other states like India, where no such regional protection exists, the role of having in place a correlative obligation to respect, protect, and fulfill a sustainable environment is of high significance.

## Conclusion

Going forward, finding a peaceful compromise between the competing interests of states and the needs of vulnerable communities is a major challenge for the future. The sovereign states have certain rights and privileges under the municipal laws; at the same time, the welfare of the people is also a duty that the states cannot negate, especially regarding climate change-induced disasters as they impact the survival, lives, and livelihood of the people that directly impact the economic, social, and cultural rights of the people. For the governments in India, both at the national and local levels, ensuring that certain safeguards regarding adaptation and mitigation are in place would fulfill the municipal responsibilities and its international obligations.

In this respect, a human rights-based approach if implemented in India will provide a method for integrating human rights within climate change policies. This would eventually lead to the integration of loss and damage suffered by vulnerable communities. It will also help integrate such issues into the design, implementation, and evaluation of policies and actions concerning climate change in India.

## References

Adger, W. N., Huq, S., Brown, K., Conway, D., & Hulme, M. (2003). Adaptation to climate change in the developing world. *Progress in Development Studies*, *3*(3), 179–195.

Agrawal, A. (2008). *The role of local institutions in adaptation to climate change*. The World Bank.

Baehr, R. (2004). The United Nations. In M. Hawkesworth & M. Kogan (Eds.), *Encyclopedia of government and politics*.

Daniel, B. (2010). Introduction: Climate change and human rights: Unpacking the issues. *Georgia Journal of International & Comparative Law*, *38*(3), 511–524.

Donnell, E. (2018). At the intersection of the sacred and the legal: Rights for nature in Uttarakhand, India. *Journal of Environmental Law*, *30*(1), 135–144.

S., Duyck, S., Jodoin, & A., Johl (Eds.) (2018), *Routledge handbook of human rights and climate governance*. Routledge.

Government of India Prime Minister's Council On Climate Change, National Action Plan On Climate Change (NAPCC) 13 (2008).

Humphreys, S. (2009). Introduction: Human rights and climate change. In S. Humphreys (Ed.), *Human rights and climate change* (pp. 1–34). Cambridge University Press.

Lalit Miglani v State of Uttarakhand and Others. (2017). WPPIL 140/2015 (High Court of Uttarakhand), 64.

Lesnikowski, A., Jodoin, S., Lemay, J., Thomson, V., & Johnson, K. (2023). Human rights in climate change adaptation policies: A systematic assessment. *Climate Policy*.

Limon, M. (2009). Human rights and climate change: Constructing a case for political action. *Harvard Environmental Law Review*, *439*, 454.

Mohd Salim v State of Uttarakhand and Others. (2017). WPPIL 126/2014 (High Court of Uttarakhand). [19].

Myers, T. E. (2011). Asian values debate. In D. K. Chatterjee (Ed.), *Encyclopedia of global justice*. Springer.

Ojha, H. R., Ghate, R., Dorji, L., Shrestha, A., Paudel, D., Nightingale, A., Shreshtha, K., Watoo, M. A., & Kotru, R. (2019). Governance: Key for environmental sustainability in the Hindu Kush Himalaya. In P. Wester, A. Mishra, A. Mukherji, & A. Shrestha (Eds.), *The Hindu Kush Himalaya assessment*. Springer.

Pandve, H. T. (2009). India's national action plan on climate change. *Indian Journal of Occupational and Environmental Medicine*, *13*(1), 17–19.

Paris Agreement to the United Nations Framework. (2015, December 12). *Convention on climate change*. T.I.A.S. No. 16-1104.

Sathaye, J., Shukla, P. R., & Ravindranath, N. H. (2006). Climate change, sustainable development, and India: Global and national concerns. *Current Science*, *314*, 318–319.

Savaresi, A., & Auz, J. (2019). Climate change litigation and human rights: Pushing the boundaries. *Climate Law*, *9*(3), 244–262.

Schapper, A. (2020). From the local to the global: Learning about the adverse human rights effects of climate policies. *Environmental Politics*, *29*(4), 628–648.

Sinden, A. (2007). Climate change and human rights. *Journal of Land, Resources, & Environmental Law*, *27*(2), 255–272.

Streck, C., Keenlyside, P., & von Unger, M. (2016). The Paris agreement: A new beginning. *Journal for European Environmental & Planning Law*, *13*(1), 3–29.

Te Awa Tupua. (2017). Whanganui River Claims Settlement Act.

The World Conference on Human Rights. (1993, July 12). *Vienna Declaration and Programme of Action*, A/CONF.157/23. UN General Assembly.

Touissant, P., & Blanco, A. (2019). A human rights-based approach to loss and damage under the climate change regime. *Climate Policy*.

UN General Assembly. (1948, December 10). *Universal declaration of human rights, 217 A* (III).

UN General Assembly. (1966a, December 16). *International covenant on civil and political rights* (*Treaty series*, Vol. 999, p. 171), United Nations.

UN General Assembly. (1966b, December 16). *International covenant on economic, social and cultural rights* (*Treaty Series*, Vol. 993, p. 3). United Nations.

UN General Assembly. (1989, November 20). *Convention on the rights of the child* (*Treaty series*, Vol. 1577, p. 3). United Nations.

UN General Assembly. (1994, January 20). *United Nations framework convention on climate change: Resolution/adopted by the General Assembly*, A/RES/48/189.

UN General Assembly. (2006). A/RES/60/251.

UN General Assembly. (2007, October 2). *United Nations declaration on the rights of indigenous peoples: Resolution/adopted by the General Assembly*, A/RES/61/295.

UN Human Rights Council. (2009, January 15). *Report of the Office of the United Nations High Commissioner for Human Rights on the relationship between climate change and human rights*, A/HRC/10/61.

UNFCCC Conference of Parties, Decision 1/CP.16, The Cancun Agreements: Outcome of the Work of the Ad Hoc Working Group on Long-term Cooperative Action under the Convention, UN Doc FCCC/CP/2010/7/Add.1 (March 15, 2011), 4.

Van Der Steen, M., Scherpenisse, J., & Van Twist, M. (2018). Anticipating surprise: The case of the early warning system of Rijkswaterstaat in The Netherlands. *Policy and Society*, *37*(4), 473–490.

Verma, S., Sharma, A., Yadava, P. K., Gupta, P., Singh, J., & Payra, S. (2022). Rapid flash flood calamity in Chamoli, Uttarakhand region during Feb 2021: An analysis based on satellite data. *Natural Hazards*, *112*, 1379–1393.

Wewerinke, M. (2013). *Climate change, human rights and the international legal order: The role of the Human Rights Council*. University of Cambridge.

Chapter 7

# Climate Change From an Education Perspective: The Role of Teachers in Raising Public Awareness

*Kaan Batı*

Hacettepe University, Turkey

## Abstract

It is quite obvious that the negative consequences of climate change will be one of the most important agendas in the near future. Many societies are developing various policies and trying to raise awareness in order to be less affected by these negativities. One of the most important keys to social awareness is undoubtedly teachers. Teachers' knowledge and awareness levels about climate change can affect the attitudes of future generations toward the causes and consequences of climate change. For this reason, in this chapter, I discuss climate change and its possible problems, the effects of climate change in Turkey and the policies carried out, public awareness of climate change, climate change awareness of teachers in Turkey and teacher education in the context of climate change. I believe that the discussions in this chapter can shed light on education policies.

*Keywords*: Climate change; sustainable development education; public awareness; teachers; teachers' perceptions

## Climate Change: Putting a Spotlight on the Problem

According to the scenarios put forward by the United Nations Intergovernmental Panel on Climate Change (1PCC), the global temperature is projected to increase by an average of 1–3.5 degrees by 2100 (IPCC, 2001). The consequences of this change carry the risk of bringing about many adverse events such as floods, overflows, droughts and ultimately desertification, storms and even epidemics that reach disaster dimensions due to situations such as sea-level rise, change in

---

**The Social Consequences of Climate Change, 79–89**

**Published under exclusive licence by Emerald Publishing Limited**
**doi:10.1108/978-1-83797-677-520241007**

temperature and precipitation regimes (Öztürk, 2002). There is a consensus that the main cause of these possible scenarios is global warming caused by the increase in human-induced greenhouse gas emissions (Öztürk, 2002). In the 5th Assessment Report of the Intergovernmental Panel on Climate Change (IPCC) published in 2014, it was emphasized for the first time that the cause of climate change is human activities with 95% certainty, and it was stated again in the Working Group I Report of the 6th Assessment Report of the IPCC published in August 2021. According to the report, the climate change experienced in the current era is defined as "anthropogenic (human-induced)" climate change (IPCC, 2014).

Climates, which have exhibited a dynamic and variable structure since the formation of the Earth, have experienced radical changes at the global level many times. However, it is now quite clear that the main cause of today's climate change is human activities (Kadıoğlu, 2012). According to the United Nations Framework Convention on Climate Change (UNFCCC), climate change means a change in climate as a result of human activities that directly or indirectly disrupt the composition of the global atmosphere, in addition to natural climate change observed over comparable time periods (Tompkins & Amundsen, 2008). The concentrations of greenhouse gas emissions such as Carbon Dioxide ($CO_2$), Methane ($CH_4$), Diazot Monoxide ($N_2O$) in the atmosphere have been increasing significantly since the industrial revolution due to fossil fuels, population growth, urbanization, agriculture and animal husbandry activities, deforestation, cement production and wastes (Türkeş, 2001).

75% of carbon dioxide ($CO_2$), one of the greenhouse gas emissions, originates from activities carried out in cities such as transport, buildings, industrial sector and fossil fuel-based energy production and consumption for these sectors (UNEP, 2022). Increasing $CO_2$ emission in the atmosphere causes the global average temperature to increase. According to the World Meteorological Organization (WMO), 2021 is one of the seven hottest years on record and the seventh consecutive year in which the global average temperature exceeded 1°C compared to the preindustrial period. Due to greenhouse gases, which are the source of this abnormal situation in the atmosphere, the negative effects of climate change will continue to be seen in the long term. As of 2022, the global average temperature has increased by 1.1°C compared to the preindustrial period (WMO, 2022).

In the 2022 Global Risks Report prepared by the World Economic Forum (WEF), (1) failure in climate action, (2) extreme weather events, and (3) loss of biodiversity are considered as the top three most important risks that may be experienced due to climate change (WEF, 2022). For this reason, it has become inevitable to take effective measures against these negative consequences and risks in the fight against climate change and to raise public awareness in order to minimize the damages that may be experienced. Cities should be at the center of the measures to be taken at the regional level. Because cities, with their crowded populations and the economic activities they contain, are the places that contribute the most to these negative consequences and are most affected by them. Making cities resilient to global climate change is also addressed within the scope of the UN 2030 Agenda and Sustainable Development Goals (SDGs) and has become one of the main determinants in achieving sustainable development goals (Gu, 2019). Raising awareness of the population living in cities is therefore one of the first steps that can offer the first and fastest improvement. The role of education and teachers in raising the awareness of the society is clear, and within the scope of this section, implications for how teachers can effectively fulfill this responsibility are presented. However, before that, the impacts of climate change and the policies carried out in the case of Turkey are exemplified.

## Climate Change and Policies in Turkey

Unfortunately, Turkey is among the countries most affected by climate change. In addition, it is a country that will be most affected by many negativities that may occur later and may experience major problems (Öztürk, 2002). According to the IPCC's Special Report on Climate Change and Land (2019), due to changes in the frequency, duration, and intensity of heat waves, precipitation increases in some regions, while drought is more frequent and intense in some places (including the Mediterranean, Western Asia, many parts of South America, most of Africa and Northeast Asia) (IPCC, 2019). Turkey is one of the countries facing the risk of drought due to its geographical location. The most water-intensive area in Turkey is agriculture with 73%, and 97% of the agricultural land in Turkey still uses surface irrigation.

The Food and Agriculture Organization's State of Food Security and Nutrition in the World report states that extraordinary climate events triggered by climate change put natural resource reserves, yields from agricultural production and food security at risk (IPCC, 2019). This problem directly affects not only the disruptions in the production, consumption and distribution cycle of food, but also the income sources of farmers living in rural areas, who play a leading role in the production of food on a global scale (Demirbaş & Aydın, 2020). A possible drought may not only cause a decrease in agricultural yields, but may also negatively affect energy production. In 2014, it was reported that the low rainfall in Turkey caused a 40% decrease in the energy produced in Hydroelectric Power Plants (HPP) compared to 2013 (Demirbaş & Aydin, 2020).

In Turkey, the Ministry of Environment, Urbanization and Climate Change is in charge of coordinating between ministries. Finally, with the "Capacity Building and Monitoring Project for the Realization of the Greenhouse Gas National Contribution Target," which was initiated by Istanbul Medipol University Climate Change Center (İKLİMER) in 2017 and completed in 2021, the foundations of the Republic of Turkey's efforts to combat and adapt to climate change until 2050 were determined by taking into account the previous reports, protocols and documents of all public institutions and organizations and relevant ministries related to energy, industry, agriculture, transportation, construction, waste, economy. The most important goal of the "Capacity Building and Monitoring Project for the Realization of the Greenhouse Gas National Contribution Target" carried out by Istanbul Medipol University under the supervision of the Ministry of Environment and Urbanization is to put forward the extension (projection) and sectoral analyses of Greenhouse Gas Emissions (SGS) by planning to fulfill the obligations of the Republic of Turkey in accordance with the United Nations Framework Convention on Climate Change (UNFCCC) and the Kyoto Convention with the principles set out in the National Goodwill Document (Şen, 2022).

## Public Awareness

After 2050, it is predicted that approximately 68% of the total population in the world will live in cities (UN, 2018). Increasing the climate change awareness of the population living in cities is very important in reducing the negativities that may be experienced and in determining and realizing the policies to be developed to combat climate change (Arı & Aydın, 2019). In the literature, there are studies to determine the perception of climate change risk (Korkmaz, 2018; Tvinnereim & Austgulen, 2014) or the perception of climate change adaptation and mitigation (Brügger et al., 2015; Capstick et al., 2015). IPCC states that those who will be most affected by the problems caused by climate change are cities and those living in cities. For this reason, it is necessary to increase the awareness of societies about adaptation to climate change (Anguelovski et al., 2016). Because increasing the climate change awareness of societies will help both in the formulation of climate policies and in individual responses to the causes of climate change (Yu et al., 2013).

The results obtained from the studies show that climate change awareness is affected by factors such as age, education level, income level, gender, ethnic and cultural background, political orientation, media influence, personal experience, and personal knowledge (Brügger et al., 2015; Capstick et al., 2015; Shi et al., 2016). On the other hand, the findings regarding the level of these effects differ according to the studies. For example, Kennedy (2015) found that as the level of welfare increases, the awareness of societies on climate change also increases, while Dunlap and Mertig (1995) concluded that people at low socioeconomic levels have higher environmental perceptions and awareness. On the other hand, there are also studies revealing that there is no relationship between socioeconomic level and climate change awareness (Korkmaz, 2018). Another variable that determines the climate change awareness of societies is the level of education. Studies show that people with higher levels of education are more likely to understand reports and scientific evidence on climate change (Etkin & Ho, 2007).

Considering the gender variable, it is seen that there is no clear distinction regarding climate change awareness. Especially in economically developing countries, there are studies revealing that women are more affected by the consequences of climate change and that they are more concerned about climate change impacts than men for various reasons (Crona et al., 2013). Similarly, Ağıralan and Sadioğlu (2021) reported in their research in Istanbul that women's awareness of climate change is higher than men (Ağıralan & Sadioğlu, 2021). In contrast to this situation, some studies have found that although men have a lower risk perception than women, this difference is not statistically significant

(Sullivan & White, 2019). Huda (2013) revealed that men's awareness of climate change is higher than women (Ağıralan & Sadioğlu, 2021). Individuals' awareness of climate change generally increases with age (Erol & Gezer, 2006). Karabulut (2023) explained the reason for this increase as the increase in concern for the future with age and the fact that environmental awareness is more visible in parallel with this. In addition, he stated that increasing knowledge and experience with age can support environmental awareness.

## Teachers' Perceptions and Beliefs in Türkiye

Teachers have the greatest responsibility in raising social awareness on climate change. Increasing individual awareness of climate change by schools will support individuals, especially those living in cities, to be more sensitive to the causes of climate change and thus reduce the effects of climate change. However, firstly, teachers' perceptions and beliefs about climate change should be determined. Akbulut and Kaya (2020) conducted a comprehensive study to examine the climate change awareness of primary school teachers and reported that teachers' awareness of climate change was high. The analyses revealed a moderately significant relationship between the variables of knowledge and attitude toward climate change (Akbulut & Kaya, 2020). In addition, research findings show that teachers' awareness of climate change increases with increasing age, and female teachers have higher awareness than male teachers (Akbulut & Kaya, 2020). A similar study was conducted by Karabulut (2023) with the participation of 271 teachers. The findings of the study are in parallel with the findings of Akbulut and Kaya (2020) and the awareness levels of teachers toward global climate change were found to be high (Karabulut, 2023). Similarly, in this study, it was found that the awareness level of teachers increased significantly with age (Karabulut, 2023). This finding is consistent with the discussions under the title of social awareness.

## Climate Change and Teacher Education

Especially developing countries need to produce strategies that strengthen the relationship between education, environment and natural resources for sustainable development (Zallé, 2019). Because sustainable development should be the focal point of society and social development (Fiorino, 2010). This approach is an important opportunity for future generations to realize sustainable development goals (Rosenbaum, 2017). As stated above, one of the key roles in improving the climate change awareness of societies is teachers. Therefore, universities that undertake teacher education should also share this responsibility (Boni & Walker, 2016). However, it is emphasized in the literature that there is not enough research on how climate change and sustainable development can be integrated into university programs (Albareda-Tiana et al., 2019). This situation both makes it difficult for universities to support sustainable development goals at the global level (Baena-Morales et al., 2022) and stands as an obstacle in the development of social awareness.

Higher education institutions are one of the educational institutions responsible for creating a sustainable future (Wright & Horst, 2013) and should contribute to sustainable development through their education, research and industry collaborations (Bayuo et al., 2020). De Leeuw et al. (2014) emphasized that it is essential to enable students to think and act in order to design economically, socially and environmentally sustainable futures through education for sustainable development (ESD). This is because education can contribute to improving people's thinking in favor of sustainable environmental, economic and social actions (De Leeuw et al., 2014). In the literature, it is stated that many universities around the world are making progress to support the achievement of the goals set out in the 2030 Agenda and are taking initiatives to strengthen cooperation with other institutions (Leal Filho et al., 2021). It is very important that these efforts are also made by higher education institutions in our country and that sustainable development goals are disseminated nationally.

## Sustainable Development Education

Sustainable development education is an approach that aims to provide individuals with a sustainable lifestyle and to realize social transformation by providing individuals with knowledge, attitudes and awareness about how sustainable development can be carried out (Öztürk, 2017). For this purpose, it is seen that many practices such as establishing education policies related to sustainable development, making necessary arrangements in curricula at all levels of education and planning education and training activities are tried to be implemented all over the world. Undoubtedly, sustainable development education is one of the most important elements in achieving our country's goals related to sustainable development. In this context, the Turkey Sustainable Development Report (Ministry of Development, 2012) prepared for the United Nations Conference on Sustainable Development (Rio+20) emphasized the need to increase the knowledge and awareness of future generations about the environment and sustainability through the inclusion of sustainable development in education programs.

Following the determination of this goal, it would not be wrong to say that research on sustainable development education has started to gain momentum in our country. When the literature is examined, it is seen that the studies focusing on sustainable development education, especially in the last decade, have gained intensity in social studies, geography and science courses and are generally studies aimed at determining the knowledge, attitudes and awareness of preservice teachers about sustainability (Akgül & Aydoğdu, 2020). Of course, one of the important pillars of sustainable development education is teacher education, and it is seen that sustainable development education in our country has been given more weight in recent years, especially at the undergraduate science teacher education level (Nas & Çoruhlu, 2017).

When the researches were examined, it was observed that the knowledge, values, and attitudes of preservice primary education and science education teachers about environment and sustainability issues increased throughout their university education (Faiz & Bozdemir Yüzbaşıoğlu, 2019). This situation can be seen as an opportunity because it can be inferred that preservice teachers do not have a negative opinion or prejudice about the environment and sustainability issues, and that they have knowledge deficiencies that can be overcome with a planned teaching activity. In this context, preservice teachers' knowledge and sensitivity about environment and sustainability issues can be increased through applied courses in primary education and science education undergraduate programs. However, there is a need to increase research on how preservice teachers can transfer their knowledge, attitudes, and awareness to their students in service as well as determining their knowledge, attitudes, and awareness.

In our country, the inclusion of sustainable development issues in the curricula started with the 2005 science curriculum. The emphasis on sustainable development was continued in the subsequent 2013 and 2018 curricula. One of the aims of the most recent 2018 science curriculum is stated as follows: "To make students realize the mutual interaction between the individual, society and the environment; to develop awareness of sustainable development regarding society, economy and natural resources" (MoNE, 2018, p. 9). On the other hand, Demir and Atasoy (2021) revealed in their research that most of the students were not aware of the concept of sustainable development (83.3%). This finding points to a very serious problem, because although it has been included in the curricula for many years, it is seen that the vast majority of students do not even have knowledge about sustainable development, let alone developing attitudes toward it. Following this situation determination, possible solution suggestions have been put forward by some researchers.

According to Öztürk (2017), sustainable development education practices in basic education require individual or institutional efforts of teachers and school administrators. Sustainable development education practices can be directly included in education programs or can be planned as extracurricular activities that can be carried out independently of education programs. It is clear that school administrations need to take planning and initiative for extracurricular practices. Öztürk also emphasized that sustainable development education should not be limited to the field of science education, and that planning should be carried out to cover other fields as well. For this purpose, all branch teachers can integrate sustainable development education practices into their annual plans in the plans prepared at the beginning of the academic year. Focusing on sustainable development education practices in students' performance and project studies can provide permanent learning experiences by allowing students to develop their critical thinking, problem-solving, and creativity skills more.

Teksöz (2016), on the other hand, suggested that in order for sustainable development education to be carried out in a healthy manner, priorities for sustainable development education should be determined, and in line with these priorities, priority project topics should be determined and managed by a center. This center should be established with the participation of the Ministry of

National Education (MoNE), the Ministry of Environment and Urbanization, the Ministry of Water and Forestry, UNESCO Turkey, local administrations, nongovernmental organizations and universities. He stated that the projects planned according to the above-mentioned sustainable development education priorities should be monitored by this center and a measurement and evaluation system should be established and operated. In order to involve all other stakeholders in the process, formal schools should be supported to prepare their own strategies in parallel with this plan. The sustainable development education strategies to be developed by schools should be carried out in cooperation with local administrations, nongovernmental organizations, and regional universities. In addition, it was emphasized that good examples should be rewarded and international exchange and participation should be strengthened.

## References


Ağıralan, E., & Sadioğlu, U. (2021). İklim değişikliği farkındalığı ve toplum bilinci: İstanbul örneği. *Anadolu Üniversitesi Sosyal Bilimler Dergisi, 21*(2), 627–654.

Akbulut, M., & Kaya, A. A. (2020). Bir afet olarak küresel iklim değişikliği ve ilkokul öğretmenlerinin iklim değişikliği farkındalığının incelenmesi: Gümüşhane İli örneği. *Gümüşhane Üniversitesi Sağlık Bilimleri Dergisi, 9*(2), 112–124.

Akgül, F. A., & Aydoğdu, M. (2020). Ortaokul öğrencileri için sürdürülebilir yaşama yönelik farkındalık ölçeğinin geliştirilmesi. *Trakya Eğitim Dergisi, 10*(2), 378–393. https://doi.org/10.24315/tred.633081

Albareda-Tiana, S., García-González, E., Jiménez-Fontana, R., & Solís-Espallargas, C. (2019). Implementing pedagogical approaches for ESD in initial teacher training at Spanish universities. *Sustainability, 11*(18), 4927.

Anguelovski, I., Shi, L., Chu, E., Gallagher, D., Goh, K., Lamb, Z., ... Teicher, H. (2016). Equity impacts of urban land use planning for climate adaptation: Critical perspectives from the Global North and South. *Journal of Planning Education and Research, 36*(3), 333–348. https://doi.org/10.1177/0739456X16645166

Arı, İ., & Aydın, L. (2019). Türkiye'de yerel iklim değişikliği eylem planlarının hazırlanması ve etkinuygulanması için öneriler. *Iğdır Üniversitesi Sosyal Bilimler Dergisi.* (Ek sayı), Erişim adresi. http://sosbilder.igdir.edu.tr/Makaleler/488895687_18-Mak-395-414.pdf

Baena-Morales, S., Prieto-Ayuso, A., Merma-Molina, G., & González-Víllora, S. (2022). Exploring physical education teachers' perceptions of sustainable development goals and education for sustainable development. *Sport, Education and Society*, 1–18.

Bayuo, B. B., Chaminade, C., & Göransson, B. (2020). Unpacking the role of universities in the emergence, development and impact of social innovations–A systematic review of the literature. *Technological Forecasting and Social Change, 155*(2020), 1–12, 120030.

Boni, A., & Walker, M. (2016). *Universities and global human development: Theoretical and empirical insights for social change* (1st ed.). Routledge. https://doi.org/10.4324/9781315742793

Brügger, A., Dessai, S., Devine-Wright, P., Morton, T., & Pidgeon, N. (2015). Psychological responses to the proximity of climate change. *Nature Climate Change*, *5*(12), 1031–1037. https://doi.org/10.1038/nclimate2760

Capstick, S., Whitmarsh, L., Poortinga, W., Pidgeon, N., & Upham, P. (2015). International trends in public perceptions of climate change over the past quarter century. *Wiley Interdisciplinary Reviews: Climate Change*, *6*(1), 35–61. Erişim adresi. https://onlinelibrary.wiley.com/doi/full/10.1002/wcc.321

Crona, B., Wutich, A., Brewis, A., & Gartin, M. (2013). Perceptions of climate change: Linking local and global perceptions through a cultural knowledge approach. *Climatic Change*, *119*(2), 519–531. https://doi.org/10.1007/s10584-013-0708-5

De Leeuw, A., Valois, P., & Seixas, R. (2014). Understanding high school students' attitude, social norm, perceived control and beliefs to develop educational interventions on sustainable development. *Procedia-Social and Behavioral Sciences*, *143*, 1200–1209.

Demir, Y., & Atasoy, E. (2021). Ortaokul öğrencilerinin sürdürülebilir kalkınmaya yönelik algılarının incelenmesi. *Trakya Eğitim Dergisi*, *11*(3), 1688–1702. https://doi.org/10.24315/tred.878404

Demirbaş, M., & Aydin, R. (2020). 21. Yüzyılın en büyük tehdidi: küresel iklim değişikliği. *Ecological Life Sciences*, *15*(4), 163–179.

Dunlap, R. E., & Mertig, A. G. (1995). Global concern for the environment: Is affluence a prerequisite? *Journal of Social Issues*, *51*, 121–137. https://doi.org/10.1111/j.1540-4560.1995.tb01351.x

Erol, G. H., & Gezer, K. (2006). Prospective elementary school teachers' attitudes toward environment and environmental problems. *International Journal of Environmental & Science Education*, *1*(1), 65–77.

Etkin, D., & Ho, E. (2007). Climate change: Perceptions and discourses of risk. *Journal of Risk Research*, *10*(5), 623–641. https://doi.org/10.1080/13669870701281462

Faiz, M., & Bozdemir Yüzbaşıoğlu, H. (2019). Öğretmen adaylarının sürdürülebilir kalkınma farkındalıkları. *Bolu Abant İzzet Baysal Üniversitesi Eğitim Fakültesi Dergisi*, *19*(4), 1255–1271.

Fiorino, D. J. (2010). Sustainability as a conceptual focus for public administration. *Public Administration Review*, *70*, 78–88.

Gu, D. (2019). Exposure and vulnerability to natural disasters for world's cities. 13 Ocak 2022 tarihinde. https://www.un.org/en/development/desa/population/publications/pdf/technical/TP2019-4.pdf adresinden erişildi

Huda, N. (2013). Understanding Indigenous people's perception on climate change and climatic hazards: A case study of Chakma Indigenous Communities in Rangamati Sadar Upazila of Rangamati District, Bangladesh. *Natural Hazards*, *65*(3), 2147–2159. https://doi.org/10.1007/s11069-012-0467-z

Intergovernmental Panel on Climate Change (IPCC). (2001). Climate change 2001: The scientific basis. In J. T., Houghton, Y., Ding, D. J., Griggs, M., Noguer, P. J., van der Linden, X., Dai, K., Maskell, & C. A., Johnson (Eds.), *Contribution of working group I to the third assessment report of the intergovernmental panel on climate change* (p. 881). Cambridge University Press.

IPCC. (2014). *Climate change 2014 synthesis report, contribution of working groups I, II and III to the fifth assessment report of the intergovernmental panel on climate change*. IPCC.

IPCC. (2019). Land: An IPCC special report on climate change. In *Desertification, land degradation, sustainable land management, food security, and greenhouse gas fluxes in terrestrial ecosystems* (Vol. 41).

Kadıoğlu, M. (2012). *Türkiye'de İklim Değişikliği Risk Yönetimi. Türkiye'nin İklim Değişikliği II.* Ulusal Bildiriminin Hazırlanması Projesi Yayını.

Karabulut, N. (2023). Öğretmenlerin küresel iklim değişikliğine yönelik farkındalıklarının değerlendirilmesi. *Ulusal Eğitim Dergisi*, *3*(2), 265–294.

Kennedy, B. (2015). *Describing and explaining cross-national public opinion on climate change*. (Doktora Tezi) ProQuest veri tabanından erişildi. (ProQuest numarası: 3726165).

Korkmaz, M. (2018). Public awareness and perceptions of climate change: Differences in concern about climate change in the West Mediterranean region of Turkey. *Applied Ecology and Environmental Research*, *16*(4), 4039–4050. https://doi.org/10.15666/aeer/1604_40394050

Leal Filho, W., Frankenberger, F., Salvia, A. L., Azeiteiro, U., Alves, F., Castro, P., .., & Avila, L. V. (2021). A framework for the implementation of the Sustainable Development Goals in university programmes. *Journal of Cleaner Production*, *299*, 126915.

Ministry of Development. (2012). *Turkey's sustainable development report: Claiming the future 2012*. http://www.surdurulebilirkalkinma.gov.tr/wp-content/uploads/2016/07/1.Claiming_the_Future.pdf

Ministry of National Education (MoNE). (2018). Fen bilimleri dersi öğretim programı [Science Curriculum]. https://mufredat.meb.gov.tr/Dosyalar/201812312311937-FEN%20BİLİMLERİ%20ÖĞRETİM%20PROGRAMI2018.pdf

Nas, E. R., & Çoruhlu, T. Ş. (2017). Fen bilgisi öğretmen adaylarının perspektifinden sürdürülebilir kalkınma kavramı. *Yüzüncü Yıl Üniversitesi Eğitim Fakültesi Dergisi*, *14*(1), 562–580.

Öztürk, K. (2002). Küresel iklim değişikliği ve Türkiyeye olası etkileri. *Gazi Üniversitesi Gazi Eğitim Fakültesi Dergisi*, *22*(1), 47–65.

Öztürk, M. (2017). Sürdürülebilir gelişme odaklı eğitim: kuramsal çerçeve, tarihsel gelişim ve uygulamaya dönük öneriler. *Ilköğretim Online*, *16*(4), 1–11. https://doi.org/10.17051/ilkonline.2017.342997

Rosenbaum, E. (2017). Green growth—Magic bullet or damp squib? *Sustainability*, *9*(7), 1092.

Şen, Z. (2022). İklim Değişikliği ve Türkiye. *Çevre, Şehir ve İklim Dergisi*. Yıl: 1. Sayı: 1. ss, 1–19.

Shi, J., Visschers, V. H., Siegrist, M., & Arvai, J. (2016). Knowledge as a driver of public perceptions about climate change reassessed. *Nature Climate Change*, *6*(8), 759–762. https://doi.org/10.1038/nclimate29972015GL064458

Sullivan, A., & White, D. D. (2019). An assessment of public perceptions of climate change risk in three western U.S. Cities. *Weather, Climate and Society*, *11*(2), 449–463. https://doi.org/10.1175/WCAS-D-18-0068.1

Teksöz, G. (2016). Geçmişten ders almak: Sürdürülebilir kalkınma için eğitim. *Boğaziçi Üniversitesi Eğitim Dergisi*, *31*(2), 73–97.

Tompkins, E. L., & Amundsen, H. (2008). Perceptions of the effectiveness of the United Nations Framework Convention on Climate Change in advancing national action on climate change. *Environmental Science & Policy*, *11*(1), 1–13.

Türkeş, M. (2001). Küresel İklimin Korunması, İklim Değişikliği Çerçeve Sözleşmesi ve Türkiye. Tesisat Mühendisliği, TMMOB Makina Mühendisleri Odası. *Süreli Teknik Yayın*, *61*(ss), 14–29.

Tvinnereim, E., & Austgulen, M. H. (2014). Når det snakkes om klima, bør vi se oljearbeideren i rommet. In E. Eide, D. Elgesem, S. Gloppen, & L. Rakner (Eds.), *Klima, medier og politikk* (S. 317–334) içinde. Erişim adresi. https://www.researchgate.net/publication/292144015_Nar_det_snakkes_om_klima_bor_vi_se_oljearbeideren_i_rommet

UNEP. (2022). Cities and climate change. 14 Ocak 2022 tarihinde. https://www. unep.org/explore-topics/resource-efficiency/what-we-do/cities/cities-and-climate-change adresinden erişildi

United Nations (UN). (2018). *World urbanization prospects 2018 – Our future is urban* (Teknik rapor). United Nations web sayfasından erişildi. https://www.un.org/development/desa/publications/2018-revisionof-world-urbanization-prospects.html

WEF. (2022). *The global risks report 2022*. WEF.

WMO. (2022). 2021 one of the seven warmest years on record. WMO consolidated data shows, 14. Ocak 2022 tarihinde. https://public.wmo.int/en/media/press-release/2021-one-of-seven-warmest-years-record-wmo-consolidated-data-showsadresinden erişildi

Wright, T., & Horst, N. (2013). Exploring the ambiguity: What faculty leaders really think of sustainability in higher education. *International Journal of Sustainability in Higher Education*, *14*(2), 209–227.

Yu, H., Wang, B., Zhang, Y. J., Wang, S., & Wei, Y. M. (2013). Public perception of climate change in China: Results from the questionnaire survey. *Natural Hazards*, *69*(1), 459–472. https://doi.org/10.1007/s11069-013-0711-1

Zallé, O. (2019). Natural resources and economic growth in Africa: The role of institutional quality and human capital. *Resources Policy*, *62*, 616–624.

Chapter 8

# A New Cold War Is Brewing: Equity and Vulnerability in a Warming World

*Ellen Wasylina*

President and Founder, Trocadéro Forum Institute, France

## Abstract

The premise of this article is to demonstrate that there is a cost to getting to net zero, and this chapter will demonstrate that those that will succeed will necessarily have invested in R&D, human capital, and will have the technology to decarbonize or reduce the carbon emissions in their economy. We will look at different benchmarks that will illustrate and validate the profile of those economies which can succeed not only in terms of getting to net zero but also to attract the best talent and create new jobs. The capacity to innovate will be key, as will the ability to attract and obtain investments, or Foreign Direct Investment (FDI). Are all the political and economic conditions reunited to receive such investments?

*Keywords*: Net zero; R&D; human capital; decarbonisation; investment; innovation; energy war; technology war; chip war; artificial intelligence (AI); job destruction and creation

Everyone is equal in the face of climate change, but those countries which are in hot, desert climates, and those living in the rural South, where the income gap is the highest with the cooler, richer North, this is where we will find the most vulnerable populations, unable to farm, provide food, have access to water or fertilizers, which don't have the necessary infrastructure or the technology to help them face a warming world.

This is the new Cold War, or an Energy or Technology war that is literally heating up between countries that have invested in R&D, the research necessary to foresee and prepare for a warming world, and those countries which have not.

**The Social Consequences of Climate Change, 91–98**

**Published under exclusive licence by Emerald Publishing Limited**
**doi:10.1108/978-1-83797-677-520241008**

However, the cost and benefits do not add up:

Based on the latest cost estimates of emission reductions from the UN climate panel, he finds that fully delivering on the 1.5° Paris promise will cost 4.5% of global GDP/year by midcentury, 5.5% by 2100.

Averaged across the century, delivering the Paris climate promises would create benefits worth $4.5 trillion (in 2023 dollars) annually. That's dramatically smaller than the $27 trillion annual cost that Paris promises would incur, as derived from averaging the three cost estimates from the two Climate Change Economics papers through 2100 (Lomborg, 2023).

In other words, those that can pay for the transition can reap the benefits of cleaner, healthier, and more efficient, better performing economies.

So it is and will be cause or source of geopolitical tensions between those countries that have the means to get their countries to net zero and those that do not.

Those that cannot will be left behind, and their economies, people, health, and wealth will suffer. Wars, conflicts, and the race for resources will be magnified.

And the differences are stark, as we can see on the Global Innovation Index (Neufeld, 2023), which explains why some countries are able to create the next thing, create jobs, and keep their economies humming... which keeps them competitive, which attract the most Foreign Direct Investments. Notice the ranking:

1) Switzerland, 2) United States, 3) Sweden, 4) United Kingdom, 5) the Netherlands, 6) South Korea, 7) Singapore, 8) Germany, 9) Finland, and 10) Denmark.

And a second graphic (Studt et al., 2021) which explains how these countries stay ahead, by investing a percentage of their GDP in R&D. Notice the top players on the right hand of the graph, investing as much as 5% of GDP: Israel, South Korea, Finland, Denmark, Singapore, Sweden, Japan, United States, Austria, Germany, Belgium, Switzerland, Belgium, Taiwan; and on the left axis, the number of researchers, where we can see that Finland has the most researchers, followed by Denmark and Singapore, South Korea and Norway, to name the first five.

This article will demonstrate the importance of these benchmarks to measuring the capacity of countries, and the failure of other countries, that will result in the energy, the technology, chip, and investment wars that will lead to a new Cold War.

The balance of power in the world is in a precarious position as the post–World War II order is being challenged, similar to what happened at the Congress of Vienna in 1814 and later in 1868 with the Treaty of Westphalia.

In Daniel Yergin's epic book "*The Quest*" (Kurtz, 2012), written in 2011 before the Fukushima earthquake and subsequent destruction of the Fukushima nuclear plant, he writes that electricity consumption has doubled since 1980 and would double again by 2030. He continues by writing that:

> the cost for building the new capacity to accommodate this growth between now and 2030 is currently estimated at $14 trillion - and is rising ...but that expansion is what will be required to support what could by then be a $130 trillion world economy.

If we look at where the energy producers are, we come up with a well-known map.

## The Energy War

*While the United States is currently the largest producer of shale gas at 13 million barrels/day* (Eberhart, 2023)*, other countries, such as Venezuela, Iran, Russia, Saudi Arabia, Qatar, and Canada, have large reserves of either oil or gas.*

Just two years ago, in March 2022, when I conceptualized and hosted a Green Economy Forum in Algiers, the buzz word was hydrogen, particularly green hydrogen. Now it seems to have disappeared ...giving way to AI...

The challenge for these energy producers is to produce cleaner energy, moving away from fossil fuels, or at least balancing out their fossil production, consumption with clean renewables, or lower carbon alternatives, creating a greener energy mix that can continue to sustain our economies.

As we end 2023 on the closure of COP28 in Dubai and look forward to COP29 in Baku, it is clear that there are challenges to keep our economies humming: we need energy security to be at the forefront to guarantee political, economic, and social stability.

Getting more countries on board in a collaborative approach is a noble goal, but as we saw with the COP21 in Paris, almost all of the nations of the planet pledged to reduce carbon emissions, but few of them have really implemented policies and allotted budgets, invested in technology, infrastructure, research, human capital, to achieve the goal of preventing the earth's temperatures from rising more than 1.5° Celsius. But we have already passed this mark and are at +1.8° Celsius.

Is it just an illusion, or is it really achievable?

Let's take nuclear energy, a clean energy after hydropower, that can guarantee electricity at a constant rate. However, cost per kW and the time to build a reactor that can, at scale, provide electricity for our communities are both barriers to getting to net zero emissions (NZE), particularly for developing countries.

Why are we still an extractive economy?

Uranium: do we have enough to go nuclear?

The critical mineral war: do we have access to stop producing combustion engines and go all electric? Or should we remain modest and go for hybrid vehicles?

*Do we have the minerals we need to make the batteries? Lithium or Sodium phosphate?*

Tuning in to France 24 (Owen, 2024) to listen to Mark Owen interview his guests, Ali Al Saffar at the Rockefeller Foundation and Brent Wanner from the

IEA, they discussed energy poverty, defined as consuming less than 1,000 kWh per person and the inherent necessity of having batteries to stock the energy.

Ali Al Saffar said we will not achieve the 1.5° Celsius limit if the poorer countries don't have access to transition fuels and will only use fossil fuels for electricity. The results are startling: our Earth would warm by 2.6° Celsius, double the goal set out by the COP21 at the Paris Climate Conference in 2015.

Sharing the transition technology with lower income countries, or the Global South, is key to achieving our climate targets and making the transition possible everywhere.

## The Technology War

*Can the low and middle income countries leapfrog and implement new technologies to get them to cleaner energies, without going through their development process, as we did, on the back of fossil fuels?*

The price of electric batteries has come down by 80% (Tech@coheractio.com, 2021), and many countries, France included, have decided to produce their own batteries, ensuring their supply chains, creating jobs at home and make the cars available to all segments of the population.

Looking at France, President Emmanuel Macron has started a large campaign to reindustrialize the country (Gouvernement, 2023), after all the factories were shuttered in the 1990s and 2000s, moving east to Eastern Europe, or farther East to Asia, where labor and materials are still cheaper.

Let's look at the three battery factories that are going to be built in the North of France (Le ministère de l'Économie, 2023) and will create 10,000 jobs, building two million electric vehicles by 2030[1]:

> Inaugurée en France, le mardi 30 mai 2023 dans le Pas-de-Calais, l'usine d'Automotive Cells Company (ACC), une coentreprise de Stellantis, TotalEnergies et Mercedes-Benz, est la première méga-usine de batteries lithium-ion pour véhicules électriques en France.

In addition to building battery factories, France launched the Electric social car leasing plan (Le ministère de l'Économie, 2024) in January 2024 . . .which was shut down on February 14, 2024, a victim of its success. This pilot project was to provide electric cars to 50,000 homes/families for the modest sum of €100 per month for people outside cities and €150 a month for people in cities.

---

[1]https://www.economie.gouv.fr/actualites/la-strategie-nationale-batteries-un-axe-cle-de-france-2030-sur-les-mobilites-durables Translation: Inaugurated in France on May 30, 2023, in the Pas de Calais region, the Automotive Cells Company (ACC) factory, a co-Entreprise of Stellantis, TotalEnergies and Mercedes Benz, is the first mega factory of lithium-ion batteries for electric cars in France.

Here are the conditions for this social leasing (Le ministère de l'Économie, 2024).

- The fiscal revenue of reference for those in the workforce should be no more than €15,400 a year.
- The workers should reside at least 15 km or more from their place of employment or travel at least 8,000 km per year as part of their professional activity.
- The car rental should be at least for three years.

In order for the energy transition to be successful, we need more than renewable energies, electric cars, and batteries. Artificial intelligence is supposed to unlock great potential in helping us solve some of our energy efficiency problems, helping us find innovative solutions. And it is going to create another war front between those who can acquire the technology and those who cannot.

*And Artificial Intelligence, or AI, was on everybody's lips at the World Economic Forum this year in Davos.* (Wasylina, 2024)

Now as in the 1990s and 2000s, with the advent of AI, as it was when the factories closed, people, particularly the older people still in the workforce, or even those coming in to it, require new training, new skills, to do new jobs, as some of their initial trainings or diplomas may be inadequate or may not prepare them for this new economy.

Thousands of legacy jobs, that can be replaced with the help or assistance of AI, will obliterate jobs in many sectors.

Let's look at what sectors and jobs where machines could replace humans. . .and where they can't. . . as illustrated in a recent graph published by McKinsey (Chui et al., 2016).

If we look at the right side of the column, in green, we can see that data collection (64%), data processing (69%), and predictable physical labor (78%) can be automated by newer, more sophisticated technologies. But Thomas Frey does reassure us in his article in *Forbes India* where he says that AI or robots will never replace humans but can make us more productive (Moosath, 2023).

The human capital war, or the brain drain to the highest bidder . . .or which jobs will become redundant and squeeze a lot of laborers, administrative personnel, or other work intensive jobs out of the job market?

But who will make all the chips we need to power this new technology, which is particularly energy greedy?

## The Chip War

Taiwan currently produces the lion's share of all the microchips in the world. The US CHIPS Act (White House, 2023), to bring home the production of microchips, processors, in case China decides to invade Taiwan. . .where 94% of the chips are manufactured. In addition, it will allow the US to increase its investment in R&D:

> In the mid-1960s, at the peak of the race to the moon, the federal government invested 2 percent of GDP in research and development. By 2020, that number had fallen to less than 1 percent. Economic growth and prosperity over the last 40 years has clustered in a few regions on the coasts, leaving far too many communities behind.

Even regions in the United States have been left behind, left out of the prosperity created by R&D, concentrated on the West coast or in the Ivy League schools, which have the budgets. The trickle-down theory espoused by President Ronald Regan has not reached the lower echelons of societies and instead has created swaths of communities of populations in energy, internet, or knowledge poverty, which have gravitated to more populist movements and politicians. This phenomenon is however not unique to the United States but is also occurring in many countries across the globe.

It's no longer about capturing hearts and minds... but about capturing market share!

*The investment war: how much will it cost to fund the energy transition? 27 trillion Euros?*

In an article written by Christopher Mims in the Wall Street Journal (Mims, 2023), he warns that AI will require a lot of energy and that nuclear energy, particularly fusion energy, would be needed to meet the demand of AI data centers, literally 15 nuclear power plants!

Alex de Vries, a researcher at the School of Business and Economics at the Vrije Universiteit Amsterdam, projected in October that, based on current and future sales of microchips built by Nvidia specifically for AI,

> global power usage for AI systems could ratchet up to 15 gigawatts of continuous demand. Nvidia currently has more than 90% of the market share for AI-specific chips in data centers, making its output a proxy for power use by the industry as a whole.
>
> Digiconomist (2023)

So how are the big investors and innovators of this technology, like Microsoft and Google, going to reach their pledge of reaching net zero carbon emissions in their operations? Are they going to have renewable energies powering their data centers?

Maybe they could get inspiration from the United Kingdom where sewage wastewater could be a source of « clean heat »?

Who knew ...that wastewater could heat our communities and provide more than 20 TWh of heat annually, providing enough heat and hot water to 1.6 million homes (Turns, 2024)!

In conclusion, we will all have to invest and get creative and collaborative to finding the solutions and funding the energy transition, leaving no one behind.

## References

Chui, M., Manyika, J., & Miremadi, M. (2016, July 8). *Where machines could replace humans-and where they can't (yet)*. McKinsey & Company. https://www.mckinsey.com/capabilities/mckinsey-digital/our-insights/where-machines-could-replace-humans-and-where-they-cant-yet

Digiconomist. (2023, October 10). Powering AI could use as much electricity as a small country. https://digiconomist.net/powering-ai-could-use-as-much-electricity-as-a-small-country/

Eberhart, D. (2023, August 18). Why America's shale boom is not over. *Forbes*. https://www.forbes.com/sites/daneberhart/2023/08/16/why-americas-shale-boom-is-not-over/

Gouvernement, F. (2023, May 12). *Accélérer La réindustrialisation de la France*. info.gouv.fr. https://www.gouvernement.fr/actualite/accelerer-la-reindustrialisation-de-la-france

Kurtz, L. (2012). The quest: Energy, security, and the remaking of the Modern World, by Daniel Yergin. *Quantitative Finance*, *12*(12), 1787–1789. https://doi.org/10.1080/14697688.2012.719735

Le ministère de l'Économie. (2023, May 30). La Stratégie Nationale « Batteries »: Un axe clé de france ... https://www.economie.gouv.fr/actualites/la-strategie-nationale-batteries-un-axe-cle-de-france-2030-sur-les-mobilites-durables

Le ministère de l'Économie. (2024, February 14). Voitures électriques: Fin de la location à 100 euros par mois. https://www.economie.gouv.fr/actualites/leasing-social-location-voitures-electriques-100-euros

Lomborg, B. (2023, November 29). "Net zero" fails the cost-benefit test. https://www.wsj.com/articles/net-zero-fails-the-cost-benefit-test-paris-climate-accord-cop28-748ae52d

Mims, C. (2023, December 15). Ai is ravenous for energy. Can it be satisfied? *WSJ*. https://www.wsj.com/tech/ai/ai-energy-consumption-fc79d94f

Moosath, A. (2023, August 2). Ai, robots will never replace humans but can make us more productive and efficient: Thomas Frey. *Forbes India*. https://www.forbesindia.com/article/mentors-and-mavens/ai-robots-will-never-replace-humans-but-can-make-us-more-productive-and-efficient-thomas-frey/87211/1

Neufeld, D. (2023, January 11). Mapped: The most innovative countries in the world in 2022. *Visual Capitalist*. https://www.visualcapitalist.com/most-innovative-countries-2022/

Owen, M. (2024, February 28). We need to make sure clean energy transitions are just and are for everyone, countries and people. *France 24*. https://f24.my/A9uE

Studt, T., Delfino, J., Content, S. & Hall, H. (2021, January 11). Research & Development World. https://www.rdworldonline.com/

Tech@coheractio.com. (2021, November 23). Voitures électriques: Le Prix des batteries a chuté de 80 % en 6 ANS – avere-france. *Avere*. https://www.avere-france.org/voitures-electriques-le-prix-des-batteries-a-chute-de-80-en-6-ans/

Turns, A. (2024, January 5). The neglected clean heat we flush down the drains. *BBC News*. https://www.bbc.com/future/article/20240103-sewage-a-low-cost-low-carbon-way-to-warm-homes

Wasylina, E. (2024, January 29). Davos debrief #1 with PDIE Group: Climate transition and AI by Mediterranean Sustainability Partners. *Spotify for Podcasters*.

https://podcasters.spotify.com/pod/show/mediterranean-sustainable/episodes/Davos-Debrief-1-with-PDIE-Group–Climate-transition-and-AI-e2f0q86
White House. (2023, February 3). Fact sheet: Chips and science act will lower costs, create jobs, strengthen supply chains, and Counter China. *The White House*. https://www.whitehouse.gov/briefing-room/statements-releases/2022/08/09/fact-sheet-chips-and-science-act-will-lower-costs-create-jobs-strengthen-supply-chains-and-counter-china/

Part 3

# Future with Warming World or Is There a Hope?

## Chapter 9

# Climate Care as a New Purpose of the United Nations

*Vesselin Popovski*[a,b]

[a]O.P. Jindal Global University, India
[b]Sofia University, Bulgaria

**Abstract**

Climate change, a direct result of human activities, such as coal mining, deforestation, fossil fuels' industries, etc., has become the most pressing global challenge of the 21st century. While the environmental impact of climate change is widely recognized, its profound social consequences are equally significant. This chapter delves into the multifaceted consequences of climate change, exploring its effects on vulnerable communities, public health, migration practices, economic disparities, political stability, etc. The chapter then develops the argument that the need to combat climate change and to pay attention to its multifaceted social ramifications should be formulated as a new and separate purpose of the United Nations Organization, among other purposes, such as maintenance of international peace and security, international co-operation to ensure economic and social development, respect, and protection of human rights.

*Keywords*: Climate change; consequences; global governance; United Nations purposes; UN Charter review

## The Warmest Weather on Record

The year 2023 was the warmest year ever since the introduction of regular air temperature measurement: it was on average 0.52C warmer than 1991–2020 average, 1.15C warmer than the entire 20th century average, and 1.32C warmer than the preindustrial (1850–1990) average (NOAA Climate.gov, 2024). Every single day in the second half of 2023 had a higher global average air temperature, compared with the same day in 2022. The Copernicus Climate Change Service reported that in 2023

**The Social Consequences of Climate Change, 101–111**

**Published under exclusive licence by Emerald Publishing Limited**
**doi:10.1108/978-1-83797-677-520241009**

"climate records tumbled like dominoes" (Mooney et al., 2024), and every single day was at least with 1C warmer than the preindustrial average (The Copernicus Climate Change Service, 2023). Sea surface temperatures also overpassed previous highs (The Copernicus Climate Change Service, 2023). The year 2023 was indeed exceptional. These global warming records brought the world already much closer to breaching the internationally agreed climate maximum temperature rise target of 1.5C, as established in the 2015 Paris Agreement on Climate Change (Popovski, 2018). In the latest estimations, this can happen already as soon as in 2027 (Harvey, 2023).

The rising temperatures are not only meteorological statistics; these were felt worldwide by billions of people. There were intense heatwaves in all continents, severe wildfires in Brazil, Canada, the United States, prolonged cycles of intense droughts, followed paradoxically by similarly intense flooding in Europe, Africa, and Asia. Such extreme weather events were known in the past, but these were never experienced with such dramatically rising intensity and frequency. The extreme weather events destroy lives and livelihoods; drought turned to deluge and displaced many thousands in the Greater Horn of Africa, while deadly flooding in Libya killed more than 10,000 people (Global Climate Change, 2024). Drought affected many parts of South America, impacting transport and energy. Rapidly intensifying tropical cyclones caused devastation in coastal communities all over the world (World Meteorological Organization, 2023).

## Social Consequences of Climate Change

The interconnected social consequences of climate change help to apprehend the urgency of addressing, mitigating, and adapting to climate change, as a complex issue that requires comprehensive solutions. One of the well-established consequences of climate change is its disproportionate influence on vulnerable communities. Underprivileged and discriminated populations, indigenous and marginalized groups often suffer the most severe pain from environmental changes. Extreme weather events, cyclones, floods, and droughts can lead to devastating outcomes for such communities, already struggling with poverty, discrimination, and inadequate infrastructure.

It is not surprising that such communities are heavily reliant on food-producing sectors, such as agriculture and fisheries. And these are exactly those that are the most vulnerable from the effects of climate change. The dramatic rise and fall of temperatures, frequent changing precipitation, sea level rise, all these are scientifically proven to be already causing the most serious negative effects on vulnerable people and groups, leading to severe food shortages, reduction of income, increased spread of diseases. When vulnerable farmers face challenges like uncertain rainfall, fast changing temperatures, unpredictable crop growing, dissemination of pests, all these severely affect negatively not only their current livelihood, but also their longer term resilience.

A recent report by the World Bank (2023) addressed the social dimensions of climate change and argued that:

> certain social groups are particularly vulnerable to crises, for example, female-headed households, children, persons with disabilities, indigenous peoples and ethnic minorities, landless tenants, migrant workers, displaced persons, sexual and gender minorities, older people, and other socially marginalized groups. The root causes of their vulnerability lie in a combination of their geographical locations; their financial, socio-economic, cultural, and gender status; and their access to resources, services, decision-making power, and justice. The World Bank (2023)

The social fabric of vulnerable communities can be seriously disrupted as they grapple with the consequences of climate-related disasters. Flooding, draught, and fires lead to displacement and breach of social networks, to heightened stress and mental health problems. All these produce a cycle of vulnerability, worsening an already unequal distribution of resources. This cycle is difficult, if not impossible, to break, moreover, where vulnerabilities and existing social inequalities deepen, and where new inequalities may emerge, based on gender, exclusion, discrimination, psychological stress, disabilities, etc.

Another major consequence from climate change is that health problems – both physical and mental health – increase and deteriorate, due to increase in the frequency and intensity of extreme heatwaves, exposing communities to air pollution. Rising temperatures influence the spread of infectious diseases, such as malaria and dengue; these diseases expand their geographical range when extreme weather and high temperatures create favorable conditions for bacteria and viruses to spread. When diseases spread to more regions, this burdens already overwhelmed healthcare systems that become ill-equipped to handle ever growing and persistent health threats. The climate-related illnesses fall disproportionately on communities with lack of access to regular healthcare, they also amplify various existing disparities.

The climate-related diseases have been constantly on the rise, particularly in urban areas where the "urban heat island effect" exacerbates the negative consequences from the high temperatures. Vulnerable people – children, elderly, impoverished, jobless, homeless, those with pre-existing health conditions, face increased health risks. In addition to physical health, the mental health also deteriorates from the impacts of climate change, including anxiety, depression, post-traumatic stress disorders, etc. These are becoming more evident and intense as communities grapple with the loss of homes, livelihoods, sense of human security, described above as a result from extreme weather events.

Apart from health issues, another major consequence from climate change is forceful migration. Rising sea levels, land and soil degradation, deforestation, extreme weather, environmental degradation, etc., drive many people to move and may result in major and massive displacement of populations, both within and across borders. This migration, often referred to as climate migration, has significant social consequences. Internal displacement due to climate events increases social tensions, strains the capacity of receiving communities and can lead to increased competition for resources. In some cases, entire communities may need to relocate, presenting challenges of preserving cultural identities and adapting to new identities. International migration can exacerbate and jeopardize relationships between nations, raise questions related to the rights of displaced people, demand a coordinated global response to address social, economic, and political challenges associated with such migration.

The economic consequences of climate change are deeply intertwined with social dynamics, often exacerbating existing disparities and creating new challenges (Carleton & Hsiang, 2016). Sectors such as agriculture, fisheries, and tourism are particularly vulnerable to climate impacts, affecting the livelihoods of millions of people, especially in developing countries. Small-scale farmers, who often lack the resources to adapt to changing climate conditions, face declining crop yields and increased uncertainty. This can inflict a spiral of poverty, as communities dependent on agriculture struggle to maintain their way of life. In coastal areas, where many communities rely on fishing for sustenance and income, rising sea levels and ocean acidification pose existential threats to both nature and people, both ecosystems and livelihoods. Furthermore, the economic consequences of climate change can contribute to social unrest and conflict. Competition for dwindling resources, such as water and arable land, can escalate tensions within and between communities. This is particularly evident in regions already grappling with political instability and economic challenges (Mach et al., 2019).

Climate change is a recognized threat multiplier, amplifying existing vulnerabilities and contributing to political instability and security risks. The social consequences of climate change – displacement, resource scarcity, economic disparities – can strain governance structures, moreover for governments already struggling to meet the social needs of their populations and the added pressure from climate challenges can lead to unrest and political upheaval. Competition for scarce resources, particularly water and arable land, can escalate into conflicts with long-lasting social and political ramifications. The global nature of climate change requires international cooperation and coordinated efforts to mitigate its impacts. Failure to address climate challenges collectively can strain diplomatic relations, hinder cooperation on other global issues, and contribute to a more insecure and unstable world.

Another climate impact interacts with existing gender dynamics, often amplifying gender inequalities and affecting women disproportionately. Women, particularly in developing countries, are often more dependent on climate-sensitive sectors for livelihoods and are therefore more vulnerable to the effects of climate change. For example, women in agricultural communities experience increased workloads as they adapt to changing growing conditions and cope with the loss of crops. Moreover, in the aftermath of climate disasters, traditional gender roles may become more rigid, with women expected to take on caregiving responsibilities, further limiting their economic opportunities, social interactions and mobility. As communities face the challenges of climate migration, gender-based violence may increase, and women may be at a higher risk of exploitation and abuse. Recognizing and addressing the gender dimensions of climate change is crucial for developing effective and equitable responses that take into account diverse experiences and vulnerabilities of different population groups.

The urgency to address climate change is not solely an environmental imperative but a pressing political and socioeconomic factor that demands comprehensive and inclusive solutions. Vulnerable communities, often marginalized and disadvantaged, bear the brunt of climate impacts. Disruptions to agriculture and fisheries, exacerbated by extreme weather events, lead to food insecurity and economic hardships. The unequal distribution of resources and vulnerabilities creates a cycle of poverty and leaves communities ill-equipped to adapt to the rapidly changing climate. Health impacts, ranging from the spread of infectious diseases to the mental health toll of climate-related disasters, further strain communities. The unequal access to healthcare exacerbates health disparities, placing vulnerable populations at greater risk. As heat-related illnesses increase and infectious diseases expand their reach, the burden on healthcare systems becomes unsustainable. Migration, both internal and international, adds another layer of complexity to the social consequences of climate change. Displacement strains receiving communities, leading to competition for resources and potential conflicts. Preserving cultural identities and adapting to new environments become challenges for displaced populations. At the global level, climate migration tests diplomatic relations and raises questions about responsibility and rights.

Economic disparities, intertwined with climate change, perpetuate cycles of poverty, particularly in sectors highly sensitive to climate conditions. Small-scale farmers and coastal communities face existential threats to their livelihoods, contributing to economic instability. The resulting competition for resources can escalate into conflicts, threatening political stability and security. Political stability is itself at risk as governance structures grapple with the challenges posed by climate change. The strain on governments, particularly in regions already facing instability, can lead to social unrest and conflicts. The global nature of climate change necessitates international cooperation, and failure to address it collectively can have far-reaching geopolitical consequences. Gender dynamics play a significant role in the social consequences of climate change, with women often disproportionately affected. Traditional gender roles may become more rigid in the aftermath of climate-related disasters, limiting economic opportunities for women. As communities undergo climate-induced migration, women may face increased risks of gender-based violence and exploitation.

In addressing the social consequences of climate change, it is imperative to adopt a holistic and intersectional approach. Policies and interventions must consider the diverse experiences and vulnerabilities of different population groups, prioritizing the needs of the most marginalized. Recognizing the interconnectedness of social, economic, and environmental systems is essential for developing effective and equitable responses. Mitigation and adaptation strategies should prioritize building resilience in vulnerable communities, ensuring access to healthcare and education, and promoting sustainable and inclusive economic development. Moreover, international collaboration is crucial to address the global nature of climate change and its far-reaching consequences.

## Climate Care as a New Purpose of the United Nations

When the UN Charter was drafted in 1944–45, environmental protection, climate change, biodiversity loss, pollution etc., were not seen as priorities, even less as existential threats to humanity and planet. The major preoccupation at the time were the devastating effects of the Second World War and how to save future generations from these, how to reduce and eliminate armed conflicts, solve disputes peacefully, create a system of collective security, develop friendly relations among states, address economic, social, cultural issues, and respect and promote human rights and fundamental freedoms.

Eighty years later the world is significantly different. Most of the people around the world today would feel insecure not from another Third World War rather from the various extreme weather events – heavy rainfalls and flooding, devastating cyclones, extreme heats, droughts, etc.

The UN Charter needs to be revised as to add a new purpose of the UN Organization – climate care. This would address the humanity concerns with environmental degradation and climate change and focus the attention of the global institutions to solve them. Some of the nine planetary boundaries (Stockholm Resilience Centre, n.d.) have been already crossed to the point of no return; others are close to being reached. This realization has triggered mobilization among committed responsible scientists toward finding solutions. In fact many solutions already exist: there are various ways to replace fossil fuels with renewable energy sources, to reduce air pollution, to develop low-carbon industries and low–water–consuming crops, to develop effective methods of carbon capture, utilization, and storage. However, science alone won't be sufficient; there is a need for responsible good governance, both within states and internationally. The global recognition and codification of a new human right – the right to clean and healthy sustainable environment, has become an important step in this direction.

The adoption of international environmental agreements or the establishment of mechanisms to monitor and encourage international environmental protection is another important step. However, as this author pointed in a previous publication (Popovski, 2019), these are essentially voluntary agreements without legally binding enforcement power. Accordingly, the UN needs new authority to pass binding legislation and broaden its capacity to protect the global environmental system and the common goods dispute settlement mechanisms. In a world where nonstate actors including multinational corporations are major drivers of resource extraction, unsustainable production, unlimited pollution, and waste generation illustrating the tragedy of the commons, such legislation should extend to all contributors to environmental degradation. Multi–stakeholder approaches are needed involving all who are willing to contribute to a more equitable and sustainable future.

Science should be the basis for policymaking in climate change, both in terms of mitigation and adaptation. It has defined the urgency to make fundamental transformations in the many dimensions of economy, politics and human society. Only through empowered, conscientious global governance mechanisms based on science, in the spirit of true global solidarity, the hope to make the necessary changes to save our planetary home can be realized.

Suggestions for establishing new international mechanisms – such as a Global Environment Agency, a Climate Change Council (Popovski, 2020), a Global Resilience Council (Global Resilience Council, 2022), a Climate Club (Climate Club, 2024), etc., – have demonstrated the need for innovative thinking and actions.

It is encouraging to see how Vanuatu created a coalition of UN members–states requesting the International Court of Justice to issue an advisory opinion to clarify the states' obligations toward climate action and legal consequences from climate change (Vanuatu ICJ, 2023). Many courts in the world followed the precedent-setting Urgenda v. Netherlands Case (Urgenda Foundation v. State of the Netherlands, 2015) and ruled in favor of citizens' claims, urging governments to fulfill their obligations toward climate change mitigation (Sabin Centre for Climate Change Law, n.d.).

Arthur Dahl recently pointed to a serious dilemma:

> climate change is accelerating with irresistible force beyond the most alarmist scientific predictions, with a giant leap in global ocean temperatures and record heat waves around the world. To save us from a climate cataclysm, all new development of fossil fuel resources should stop immediately, and their use should be eliminated completely by 2050. Yet they are still highly subsidised, and producing countries and companies are expanding production and making enormous profits. Like a drug addiction, we cannot stop even though renewable energy sources are now cheaper than fossil fuels. No one wants to abandon proven energy sources and systems that are so profitable, and we still need technological breakthroughs for some uses and major transformations in present infrastructure and products. Arthur Dahl (2023)

Dahl argued that solution is not in going back and recreating the traditional society on a lower carbon basis, with all the other planetary limits we face, instead to use the biggest economic shift in human history to move toward greater sustainability, equality and conviviality. The UN Conference of the Parties on Climate Change, COP-28, in Dubai in November–December 2023 made a historical agreement, what can be characterized as the beginning of the end of the fossil fuel era (UNFCCC, 2023).

Similarly, the 2023 Report of the Earth Commission "*A Just World on a Safe Planet: First Study Quantifying Earth System Boundaries*" demonstrated clearly that the Club of Rome 1972 *Limits to Growth* projection was broadly right and most of the safe boundaries have been already crossed (Stockholm Resilience Centre, n.d.; Rockström et al., 2023). It made a solid scientific assessment of the state of the planetary health not only in terms of Earth System stability and resilience, but also in terms of human well-being, equity, and justice.

Both reports argued that if inequality is restrained, the majority of people can enjoy higher well-being, greater economic security, and more economic opportunities and that new models can help reverse historic injustices and open the doors for unprecedented investment in clean energy, sustainable food, and sustainable cities.

## Global Governance for Climate Change

The urgency of the climate crisis produces a chance to create a new institution, perhaps a Global Environment Agency (GEA), for those critical environmental issues of the global common good that require urgent and effective action by all states, multinational corporations, civil society and individual citizens. The UN Conference of the Parties on Climate Change, COP-28 in 2023, addressed the urgent triple planetary crisis – climate change, pollution, and biodiversity loss and took note of the 2023 State of the Climate Report, written by leading scientists, which listed 20 out of 35 vital signs of the planet have reached record extremes (Ripple et al., 2023). The Report provided further extensive evidence that the humanity is now endangering the conditions of life in our Earth system, and argued that at the center of the triple planetary crisis is the broken relationship between humanity and Nature. The view that Nature is at the service of humanity, as it has been reflected over centuries in all economic systems, has led to the depletion of life on Earth and the disruption of natural cycles through production and consumption patterns that violate the Earth's limits and break the ecological balance.

The global answer to this planetary crisis requires an urgent shift from a human-centered worldview to an Earth-centered or nonanthropocentric worldview.

Nature is finally gaining more prominence in the UN vocabulary, from the promise in the Sustainable Development Goals and the 2030 Agenda for Sustainable Development to ensure that "…economic, social and technological progress occurs in harmony with nature" (United Nations, 2015) to the plea in the 2022 Kunming-Montreal Global Biodiversity Framework, which aims to ensure that "by 2050, the shared vision of living in harmony with nature will be fulfilled" (United Nations Environment Programme, 2021).

Despite the growing global climate awareness, the UN lacks a robust mandate to implement a holistic approach and to efficiently address the deterioration of ecosystems, loss of biodiversity, and climate change. Multiple stakeholders – member states from the Global South, civil society, and academia – are advocating for the establishment and development of GEA.

What is also necessary, in a more ambitious long-term project, is to advocate for amending the UN Charter and add a new Purpose of the Organization – Climate Care. Under the Global Governance Forum, a group of scholars and activists (with this author being one of the initiators) is already working on revising the UN Charter, suggesting specific amendments (Global Governance Forum, 2023). I would argue that Article One of the UN Charter should list climate care even before the maintenance of international peace and security, economic development and human rights.

In September 2024, the UN Member States will gather in New York for the "Summit of the Future" and this is an excellent moment to revise the 80-year-old UN Charter and to add in its Preamble and Article One the new purpose of climate care. The UN General Assembly can make a single amendment to be done through already an existing provision – Art. 108 allows amendments, but also Art. 109 requires member states to review the UN Charter at any time when a simple majority of 50% plus one state see the necessity for this, without a requirement to ask a permission from the permanent members of the UN Security Council.

I would argue that without a decision to initiate a UN Charter Review Conference under Art. 109, the UN Summit of the Future in 2024 would be incomplete. What would that "Future" be, if it does not address the urgent need to include climate care in the UN Charter, along with other highly necessary changes, such as to eliminate the veto power of the permanent five members of the Security Council, empower the General Assembly with legally binding decision-making, create World Parliament, and undertake other structural reforms in the Organization.

## Conclusion

This chapter starts by expressing serious concerns with the fact that the year 2023 was the warmest year ever since the introduction of temperature measurement. Beating all existing global warming records, the world has come closer to breaching the agreed climate maximum temperature target of 1.5C rise, as established in the 2015 Paris Agreement on Climate Change and this can happen very soon. The chapter argues that the rising temperatures are not only meteorological statistics, but these have been felt worldwide by billions of people, through intense and frequent extreme weather events – heatwaves, droughts, and flooding.

The chapter describes some of the social consequences of climate change, their negative effects on vulnerable communities, heavily reliant on food-producing sectors, such as agriculture and fisheries, pointing that these are exactly those that are the most vulnerable from the effects of climate change. Health problems, including mental health issues, increase and deteriorate as a result of the extreme heatwaves, droughts, flooding produced by climate change. Millions of people can become victims of sea level rise, deforestation, land degradation and they can be forced to migrate, which can create tensions and conflicts both between communities inside states, and also between states. Economic disparities, intertwined with climate change, perpetuate cycles of poverty, particularly in sectors highly sensitive to climate conditions. The resulting competition for resources can escalate into conflicts, threatening political stability and security. Gender inequality is also a factor in the consequences of climate change, with women often disproportionately affected, traditional gender roles becoming more rigid in the aftermath of climate disasters, limiting economic opportunities for women.

The chapter assesses recent reports that state that the warnings of the Club of Rome 1972 *Limits of Growth* have materialized and recognize the dangers of crossing the nine planetary boundaries, proposing to redesign various economic and social policies to put societies on a pathway toward wellbeing for all. They suggest innovations as to drive sustainable economic growth, including for low-income countries, to accelerate change, upgrade economic systems and redefine wellbeing in economic policies. Instead of people and the planet serving the economy, the economy should serve the people and the planet.

The global nature of climate change necessitates international cooperation, and the chapter addresses proposals for strengthening existing governance structures and frameworks and creating new such. The main argument developed in the chapter is that the United Nations Charter should be revised and climate care should be included not only among its purposes but even be made as the first main purpose of the organization. The chapter ends with an appeal to convene Art. 109 Review Conference of the UN Charter at the forthcoming UN Summit of the Future in September 2024. The author is convinced that without a decision to initiate such Review Conference under Art. 109, the "Summit of the Future" may remain incomplete. What would that "Future" be, if the Summit does not initiate a review of the UN Charter to include among its purposes the most urgent need and task of humanity to survive from the climate change disasters?

## References

Carleton, T. A., & Hsiang, S. M. (2016, September 9). Social and economic impacts of climate. *Science*. https://www.science.org/doi/10.1126/science.aad9837

Climate Club. (2024). *The Climate Club*. https://climate-club.org

Dahl, A. (2023, July 21). *Climate change & world economy: When an irresistible force meets an immovable object*.

Global Climate Change. (2024, January 30). *Extreme weather and climate change*. https://climate.nasa.gov/extreme-weather/

Global Governance Forum. (2023). *A call for the UN Charter reform*. https://globalgovernanceforum.org/a-call-for-un-charter-reform/

Global Resilience Council. (2022). *A 'Security Council' for non-military threats*. https://www.foggs.org/grc-global-resilience-council/

Harvey, F. (2023, May 17). World likely to breach 1.5C climate threshold by 2027, scientists warn. *The Guardian*. https://www.theguardian.com/environment/2023/may/17/global-heating-climate-crisis-record-temperatures-wmo-research

Mach, K. J., Kraan, C. M., Adger, W. N., Buhaug, H., Burke, M., Fearon, J. D., Field, C. B., Hendrix, C. S., Maystadt, J.-F., O'Loughlin, J., Roessler, P., Scheffran, J., Schultz, K. A., & von Uexkull, N. (2019). Climate as a risk factor for armed conflict. *Nature*, *571*, 193–197. https://doi.org/10.1038/s41586-019-1300-6

Mooney, A., Bernard, S., & Bryan, K. (2024, January 9). *Climate record tumbled 'like dominoes' during world's hottest year*. Financial Times. https://www.ft.com/content/fafaf4cc-4f93-46fd-ab94-2f90165cef9f

NOAA Climate.gov. (2024). Yearly global surface temperatures compared to different long-term averages. [Graph]. https://www.climate.gov/media/15808

Popovski, V. (Ed.). (2019). *The implementation of the Paris agreement on climate change*. Routledge.

Popovski, V. (2020, June 18). *Towards multiple security councils*. Stimson Centre Issue Brief. https://www.stimson.org/2020/js2020-towards-multiple-security-councils/

Ripple, W. J., Wolf, C., Gregg, J. W., Rockström, J., Newsome, T. M., Law, B. E., Marques, L., Lenton, T. M., Xu, C., Huq, S., Simons, L., & King, D. A. (2023). The 2023 state of the climate report: Entering uncharted territory. *BioScience*, *73*(12), 841–850. https://doi.org/10.1093/biosci/biad080

Rockström, J., Gupta, J., Qin, D., Lade, S. J., Abrams, J. F., Andersen, L. S., McKay, D. I. A., Bai, X., Bala, G., Bunn, S. E., Ciobanu, D., DeClerck, F., Ebi, K., Gifford, L., Gordon, C., Hasan, S., Kanie, N., Lenton, T. M., Loriani, S., … & Zhang, X. (2023). Safe and just Earth system boundaries. *Nature*, *619*, 102-111. https://www.nature.com/articles/s41586-023-06083-8

Sabin Centre for Climate Change Law. (n.d.). *Global climate change litigation*. https://climatecasechart.com/non-us-climate-change-litigation/

Stockholm Resilience Centre. (n.d.). *Planetary boundaries*. https://www.stockholmresilience.org/research/planetary-boundaries.html

The Copernicus Climate Change Service. (2023, August 8). *Global sea surface temperature reaches a record high*. https://climate.copernicus.eu/global-sea-surface-temperature-reaches-record-high

UNFCCC. (2023, December 13). *COP28 agreement signals "beginning of the end" of the fossil fuel era*. https://unfccc.int/news/cop28-agreement-signals-beginning-of-the-end-of-the-fossil-fuel-era

United Nations. (2015). *Resolution adopted by the General Assembly on 25 September*. https://documents.un.org/doc/undoc/gen/n15/291/89/pdf/n1529189.pdf?token=vIBeaXDwqoAZdjoOpG&fe=true

United Nations Environment Programme. (2021). *Report of the conference of the parties to the Convention on Biological Diversity on its fifteenth meeting*. https://www.cbd.int/doc/decisions/cop-15/cop-15-dec-04-en.pdf.

Urgenda Foundation v. State of the Netherlands. (2015). HAZA C/09/00456689. https://climatecasechart.com/non-us-case/urgenda-foundation-v-kingdom-of-the-netherlands/

Vanuatu ICJ. (2023). *Vanuatu ICJ initiative*. https://www.vanuatuicj.com

World Bank. (2023, April 1). *Social dimensions of climate change*. https://www.worldbank.org/en/topic/social-dimensions-of-climate-change

World Meteorological Organization. (2023, December 22). *2023- A historic climate year, as WMO Secretary-General Taalas successfully completes his mandate*. https://wmo.int/news/media-centre/2023-historic-climatic-year-wmo-secretary-general-taalas-successfully-completes-his-mandate

Chapter 10

# Employing Complexity Theory in Tackling the Social Consequences of Climate Change

*Şuay Nilhan Açıkalın*

Ankarа Haci Bayram Veli University, Turkey

## Abstract

Government policy, which comprises laws, rules, regulations, standards, and incentives, is an important driver of system change. Many climate change programs, from the local to the global level, fail owing to a lack of "political will and coordination of subdimensions of public policy." Rethinking public policy is more than a requirement; it should be examined beyond the established method of government. The literature on the function of public policy is extensive; yet, the application of complexity theory (CT) to examine climate change is limited. In this regard, this chapter seeks to examine how public policy might become more effective in combating climate change by applying CT in two dimensions: reforming national policies and renewing cosmopolitanism on a local to global scale.

*Keywords*: Complexity theory; climate change; public policy; social issues; cosmopolitanism

## Introduction

Climate change is not new but is possibly the greatest threat that human social, political, and economic institutions have ever faced. Climate change is a multidimensional problem that directly influences individuals in different manners. The dangers and uncertainties originating from climate change are serious, which are paradoxically the product of our economic system. In addition to this, scientific findings about climate change are both under attack and also subject to post truth. Based on its controversial nature, the politics of climate change are harsh and difficult, it is disruptive to the psychology, and lastly environmental and nonenvironmental concerns are more complex than ever thought.

**The Social Consequences of Climate Change, 113–125**

**Published under exclusive licence by Emerald Publishing Limited**
**doi:10.1108/978-1-83797-677-520241010**

Not only the facts of climate change, which can be deemed controversial from time to time, but also the solutions are not easy to swallow. Solutions need to be addressed with both local and international efforts from all different stakeholders (Booth et al., 2020). In light of this reality, existing societal and institutional problem-solving processes were not designed, and have not developed, to deal with anything resembling an interconnected set of issues with this severity, scope, and complexity. There have been no precedents. So far, we have failed to appropriately address the situation. Problems will continue to emerge as we strive to prevent and adapt to the effects of climate change; thus, human systems must adapt (Abbass et al., 2022).

When climate change entered the domain of public policy, which was after the end of the 1980s, the World Meteorological Organization (WMO) and the United Nations Environment Programme (UNEP) founded the Intergovernmental Panel on Climate Change (IPCC) in 1988 (History of IPCC, 2024). The IPCC was tasked with assessing and summarizing the most recent scientific, technological, and socioeconomic data about climate change and publishing its conclusions in reports delivered to international organizations and national governments worldwide. Since 1988, the understanding of public policy by international organizations and governments has undergone changes based on goals in tackling climate change. Uncertainty and unpredictability have become an integral component of modern societies, with consequences for social sciences. To put it another way, social scientists discovered that human-driven events cannot be reduced merely to forecasts or an order-centered approach. Furthermore, globalization has had a huge influence on the fostering of this idea that our environment is essentially unpredictable (Urry, 2003). In the last two decades, complexity theory (CT) and its use in social sciences have been on the rise. There has been a significant accumulation of theoretical work (Bogg & Geyer, 2007; Byrne, 2002; Tomé & Açıkalın, 2019).

One key way for system change is government policy, which includes laws, rules, regulations, standards, and incentives. However, many climate change plans, from the local to the global level, fail due to a lack of "political will and coordination of subdimensions of public policy." Rethinking public policy is more than a necessity, and should be analyzed beyond the traditional way of governance. Literature about the role of public policy is quite broad; however, employing the tools of CT to revisit climate change is limited. In this respect, this chapter aims to analyze how public policy can become more efficient in tackling climate change using CT in two dimensions; reorganizing national policies and revitalizing cosmopolitanism from local to global scales.

## Climate Change: Facts, Social Consequences, and Goals

Climate change is an issue of the global commons. Its origins – man-made greenhouse gas (GHG) emissions – and consequences are dispersed and felt (although unevenly) across the international system, surpassing traditional boundaries and authorities of nations in the international political system. As a result, objectively establishing causation is very challenging. In order to understand what climate change is and how it became a vital issue for everyone, the IPCC circulated the following facts (2023):

- Human-caused global warming of 1.1 degrees Celsius has resulted in unparalleled climatic shifts in recent human history.
- Climate impacts on people and ecosystems are broader and more severe than anticipated, and future hazards will rise quickly with each fraction of a degree of warming.
- Adaptation strategies can successfully increase resilience, but more funding is required to scale solutions.
- Some climatic consequences are already too severe to adjust to, resulting in losses and damages.
- There is a greater than 50% likelihood that global temperature rises will reach or exceed 1.5 degrees Celsius (2.7 degrees Fahrenheit) between 2021 and 2040 in research.
- Burning fossil fuels is the main reason behind the climate crisis.
- Emission cuts are necessary across all of society to combat the climate crisis.
- All methods rely on some amount of carbon removal to keep warming to 1.5 degrees C (2.7 degrees F) with no or minimal overshoot.
- Today, financing flows for fossil fuels from both the public and private sectors greatly outweigh funding for climate adaptation and mitigation.
- Poorer and vulnerable populations are now and will continue to be worst harmed by the effects of climate change (IPCC, 2023: Summary for Policymakers, 2023).

These facts clearly indicate how climate change and its impacts are multidimensional. Solutions should be formulated at different levels with cooperation of various actors. In addition to this, social consequences are key, not only climate change and natural consequences. When it comes to social consequences, it should be noted that the cross-cultural perspective is important (Jaffe et al., 2019). In terms of cross-cultural reality, "historical responsibility" and "right to development" are frequently invoked in climate change arguments, although their application is restricted (Friman & Hjerpe, 2015; Mukherjee & Mustafa, 2019).

Priorities in tackling climate change have also been evolving through the years based on research. For example, in the beginning of the 2000s, priorities and goals were mostly focused on climate change while social consequences were neglected. "Characterizing climate change as a simple pollution problem may have helped focus the early policy negotiations, but things are different today" (Luers & Sklar, 2013). However, the last two decades were game changers for the social consequences of climate change, which became one of the priorities on the national and international agenda, especially after COVID-19. Also, we should note that social consequences are very diverse when it comes to facts and policymaking process. The most important feature that led us to see climate change as a new social problem is related to uncertainty (Gough et al., 2008). Accordingly, the impact of climate change is less predictable than the impact of other social problems. In other words, while the vulnerabilities due to traditional social problems such as unemployment or poverty are directly visible, problems related to climate change are diffuse, indirect, and have a significantly longer time horizon. Furthermore, Bhattacharya underlined that COVID-19 caused economic disruption in spiral of consequences, similar to climate change (2020). It is obvious that consequences of climate change can't be minimized to carbon emissions and climatic events. Health risks, food insecurity, energy geopolitics, climate justice, and migration are deemed the main topics, which involve various socio-behavioral consequences (Pongiglione & Cherlet, 2015; Safonov, 2019).

These changing patterns and trends in climate change were reflected in the goals for tackling climate change. To reduce global climate change to "safe" levels in the long run, both developing and industrialized countries must integrate climate change policies with sustainable development initiatives (Beg et al., 2002). Central concerns regarding sustainability – economic, social, and environmental – undoubtedly impact activity in all of these sectors. Preferences and priorities for each of these cross-cutting concerns impact how climate policy choices are evaluated and suggested. As expected, this chaotic and multidimensional nature of climate change points to the necessity for a holistic approach to provide solutions.

## Social Consequences of Climate Change in Public Policy: A Tough Task

As touched upon above, facts, social consequences, and goals of climate change are complex, interdependent, and dynamic. Especially social consequences and goals are hard to define, let alone to determine and suggest solutions. This is underlined in the Fourth Assessment Report by the IPCC, which defines "vulnerability" as "the degree to which geophysical, biological, and socio-economic systems are susceptible to, and unable to cope with, adverse impacts of climate change" (IPCC, 2007). In order to respond to this vulnerability, the solution seeking process within the public policy domain of climate change is also a complex process, comprising diverse subtopics that can be analyzed under two main concepts of adaptation and mitigation. Adaptation and mitigation are both distinct but integral notions that were simply defined by Laukkonen et al. (2009) as minimizing the effects of climate change (mitigation) and coping with them (adaptation).

Under the umbrella of developing public policy regarding social consequences of climate change, adaptation is an undeniable dimension. Adapting to changing climate conditions is crucial for both individuals and societies. Adaptation to climate change involves modifying natural or man-made environments. Although natural and individual adaptation seems to be feasible, individuals generally fail to comprehend the goals of tackling climate change (Clar et al., 2013). This requires the necessary capacities, resources, and context variables. Governments worldwide recognize the importance of facilitating climate change adaptation through public policies, as self-adaptation is often insufficient and can lead to negative macroeconomic consequences (Cimato & Mullan, 2010). While most governments are keen to develop public policies to address the social consequences of climate change, there is limited understanding about prioritizing adaptation investments, policies, and programs. In addition to this, and more importantly, most governments lack information to combine existing social issues such as poverty, unemployment, and gender equality with adaptation policies.

The second dimension of public policy for climate change is mitigation. In order to prevent extremes of global warming, climate mitigation refers to the worldwide decrease of GHGs (carbon dioxide, methane, nitrous oxide, and sulfur hexafluoride) caused by humans. The primary type of emissions is carbon dioxide, which is created when fossil fuels are used for transportation, manufacturing, electricity generation, and agriculture (IPCC, 2007). From this perspective, Gough et al. recommend framing public policy initiatives to mitigate climate change within the context of emerging societal dangers because mitigation policies cannot be employed with only the technical perspective. Climate change poses unique and complicated dangers compared to typical societal issues. Its impact is global, long-term, and very unclear (2008).

Thus, in the fight against climate change, in which the state is the main actor, a management approach in which all stakeholders and all institutions are included becomes possible. Thus, the existence of an active state in which rights are determined, protected, and full utilization is ensured will establish fair distribution by ensuring that social policies are created for the most vulnerable individuals (Gough et al., 2008). In this sense, mitigation (energy, transport, forestry, trade, etc.) and adaptation (education, health, housing, social security, employment, etc.) policies and programs needed to combat climate change are the main focus of social policy (Memiş & Aydin, 2023). Put simply, the traditional conception of industrial societies – including their carbon-based technologies, economies, growth principles, and lifestyles – are called into question by climate change, necessitating a fundamental shift in society and culture in order to implement mitigation and adaptation strategies.

## Complexity Theory and Public Policy

Rethinking adaptation and mitigation within the public policy framework needs more complex and comprehensive perspectives to capture the social aspects. Capra suggests that understanding nonlinear processes and complex networks, as well as CT, might help to solve complicated problems (2005). CT is a nonlinear paradigm compared to the Newtonian understanding of the public policy. The second half of the 20th century saw significant advancements in postpositivist approaches to social science. The term "complexity theory" refers to natural science techniques that deal with nonlinear, complex, and chaotic systems, as opposed to Newtonian science, which focuses on path dependence and predictability (Lindblom, 1959). It is challenging to provide a concise introduction to CT due to the lack of a universally accepted concept or theory. The term's ambiguity arises from its broad scope in nonlinear dynamic systems research rather than being a specific scientific theory. Mitleton-Kelly (2000) gives a simpler definition of complexity referring to the deep interconnectedness of elements within a system and its surroundings. This description is an excellent beginning point.

Although CT emerged in the natural sciences since the 1960s, both natural scientific trends and their application to social science have been contested. The implications of CT have been diverse in the social sciences. The end of the Cold War led to the formation of new paradigms in public administration (Erkoçak & Açıkalın, 2015). In addition to this, Cairney and Geyer suggested that CT can be used as bridge between policymakers and academic perspectives to provide a more holistic perspective (2015). Aligned with this approach, Morçöl clearly suggests that CT provides techniques and concepts for comprehending the macro/micro challenges that underpin the public policy process (2012). Complexity theorists predict nonlinearity in policy processes due to the interdependence of complex systems and their ongoing interaction, evolution, and self-organization. Thus, it can be said that the new world is characterized by nonlinearity, unpredictability, and fluidity, with multiple players and levels of government, which include the main concepts of CT which are multi-actor, nonlinearity, coevolution, and self-organization (Açıkalın, 2022; Klijn, 2008; Meek, 2010). Unpredictability in policy processes and limited control by important actors stem from the interaction of variables that impact and restrict policymaking rather than their presence alone.

First of all, one feature of CT is the *complex adaptive system*. The same components – later referred to as actors – interact in various ways inside this system. In nonlinear interactions, actors form networks in which they act and respond to one another. Secondly, a *multi-actor* structure is a major source of unpredictability in a complex adaptive system due to interactions over time and space, which impact and are influenced by each other. Thirdly, the multi-actor structure forms unending positive and negative feedback loops through interactions. Nonlinear relationships between actors can lead to unorthodox rules, unlike route dependent structures. Fourthly, CT promotes *coevolution* which is "the evolution of one domain or entity (that) is partially dependent on the evolution of other related domains or entities, or one domain or entity changes in the context of the other(s)" (Mitleton-Kelly, 2003). Lastly, self-organization is a system's ability to spontaneously self-organize itself into higher levels of complexity. When the constituents of a complex system interact locally, the system as a whole reshapes and renews itself in a spontaneous response to changes in the external environment (McMillan, 2003). As touched upon, there are growing numbers of approaches and schemes in CT to be used in public policy that need to be limited (Morçöl, 2010). In this respect, for this chapter, public policymaking process is considered to be a complex adaptive system. The focus is on three notions, which were explained above, of *multi-actor, self-organization, and coevolution* in order to identify solutions for tackling the social consequences of climate change through decision-making, implementation, and evolution of public policies.

## Reconsidering Public Policy for Climate Change With CT

Developing and implementing public policies for the social consequences of climate change is a complicated process from local to international level. As drawn, linear policymaking processes for social consequences of climate change have several weaknesses: (1) a lack of interaction with the climate system, (2) ignorance of the presence and interactions of particular actors and decision-makers that provide mechanisms for collective behavior, and (3) an inability to forecast direct and indirect consequences of policies from a cross-sectional perspective (Braunerhjelm & Hepburn, 2023). Multiple actors face dynamic environmental, economical, and institutional circumstances that influence their cooperation and competition to attain procedural and substantive goals (Macintosh & Wilkinson, 2016). Although the application of CT has been extending in the subdimensions of public policy, literature about climate change and policymaking can be deemed to be limited (Mach et al., 2014; Istiadji et al., 2018; Kingdon, 1995). As mentioned in the previous section, patterns in the policymaking process can be revisited using CT to shed light on more realistic and holistic perspectives that will also positively contribute to sustainable development. In order to understand how CT can be employed in the policymaking process, once more it should be noted that social consequences of climate change here refer to a wide range of social, economic, and cultural issues which are not restricted to any specific policy area.

### *Agenda Setting and Policy Formulation*

*Multi-actor Structure:* As Dunn suggests, it is impossible to break down subjectively experienced problems like crime, poverty, unemployment, inflation, energy, pollution, health, and security into separate subgroups without taking the risk of determining a roughly correct solution to the incorrect problem. The fact that a system of problems is more than the mere sum of its components, or is qualitatively different, is one of its fundamental characteristics (Dunn, 2017). Public policy for the social consequences of climate change itself involves diverse actors such as leaders, key figures, mid-level and high-level bureaucrats, interest groups, international organizations, and civil society. In other words, complexity matters, especially when the skills and information needed to solve the issues are distributed among people without solid institutional linkages. Divergent interests, opposing aims, and competing narratives are inherent in many complicated challenges (Da Silva & Lammersen, 2017). This reality is vital to understand, especially when determining priorities in policymaking and agenda setting. Despite agenda setting in climate change in the natural sciences being under investigation, such as anticipating global climate patterns and temperature changes, our ability to downscale these global predictions to the regional scale is limited. In other words, the ability to quantify the impact of regional ecological change on social and economic institutions will allow for the quantification of regional costs based on various climate impacts that reflect agenda setting (Mach et al., 2014). CT offers more flexible ways to cope with adversity in multi-actor structures. For example, Geyer and Rihani suggested that soft management methods in agenda setting, which replace a top-down process with an inclusive perspective (2010). Furthermore, Erkoçak and Açıkalın suggested bringing actors into agenda setting with constructive elements, which can be laid down in the characteristics of meta-governance (2015). Of course, determining priorities also should have a scientific background involving fuzzy set comparative qualitative analysis and computational modeling (Carmona et al., 2023; Douglas et al., 2020). When it comes to policy formulation, for an inclusive approach to this dynamic public policy process, governments should adopt constructive elements that center the disparate priorities of different actors. In other words, while governments seek a flexible approach to agenda setting, they nudge actors to make efficient and feasible policy formulations that contribute to coping with the social consequences of climate change (2015). Lastly, policy formulation encompasses externalization and internalization within political discourse, which also highly lead to substantial changes in policy agendas, especially in climate change (Jann & Wegrich, 2003).

### *Self-Organization*

*Implementing Agenda:* Implementing policies is the second stage of the public process. In the traditional understanding of public policy, implementation of agenda is governed and controlled by a hierarchical structure, which ought to be revisited with the self-organization properties of complex adaptive systems. Implementing agenda priorities for climate change is also an unpredictable and multidimensional process, and the selected policy subsystem is only part of it. It encompasses the framing of tactics and tools used to address climate change, the management of key policy players, and the procedures used to apply policy instruments (Eliadis et al., 2005). Self-organization refers to the spontaneous generation of new forms of order, similar to the autopoiesis power response, which achieves dynamic equilibrium. A self-organizing system must also be adaptable and possess homeostatic skills to sustain its condition through internal interaction. Mason (2007) underlined that the process of self-organization is not controlled by an external party but results from spontaneous interactions between the agents within the system. For example, policy implementation in the education sector as a subsystem for the social consequences of climate change involves teachers, students and school administrations as stakeholders who will self-organize their behavior based on the new conditions in a natural way either by adapting or resisting. This process is generated mostly by stakeholders by themselves in a natural way, but later can create different results for implementation, such as bringing strict measures. In other words, the policy implementation process should be analyzed as an unpredictable reality regarding effectiveness and the potential domino effect.

### *Coevolution*

*Monitoring and Evaluation*: Policy relevance, efficiency, and effectiveness are often evaluated at the conclusion of the policy cycle (Jann & Wegrich, 2003). The assessment may result in the termination or renewal of selected policies. Especially in the case of climate change issues, evaluations should serve two main aims: benefits for emission reductions and minimizes negative effects of mitigation and adaptation policies. From the CT perspective, feedback loops through actors within the system generate continuation of policies in the correct manner. According to CT, reinforcing or positive feedback amplifies or accelerates a change away from a starting point or equilibrium point, whereas balancing or negative feedback dampens, slows down, or corrects a change in a system that is moving away from the starting point. In simple words, less leads to less or more leads to less (Kastens, 2010). Also, the multi-actor structure of the policymaking process and self-organization clearly show that the social consequences of climate change comprise different policy areas. This structure of social consequences and public policies can evolve through time. For example, public policies about mental health can create a domino effect in different policy areas such as cultural policies; therefore, the coevolution process is not only about learning and adapting to each other. In this respect, the effectiveness of one-policy area leads to emergence in different policy areas. In this respect, CT strongly highlights how policy evaluation should be multidimensional and systemic (Haynes, 2008). Similar to complex adaptive systems in nature, some actors can play facilitator roles. In order to escape policy disconnects or fragmentation after systemic-level evaluation, the civil society component plays a vital role as a bridge between actors to transmit feedback between actors and also can be used as a platform to easily adapt to new circumstances.

## Conclusion

Climate change is both one of the devastating and controversial problems that humanity faced. Facts and policymaking process regarding climate change is highly complex and dynamic. Especially social consequences of climate change have been more diverse and demand long-term policies. Mitigation and adaptation composed climate change policies in terms of content and sectorial diversity that is also interdependent with each other. As all known, public policy process has been under transformation that istriggered by characteristics of 21st-century multi-actor and unpredictability. Especially, social consequences of climate change are solely a tough task for policymakers and stakeholders. In this respect, modeling and complex systems science provides a fine-grained understanding of human activity at both individual and group levels, particularly in situations where human and environmental systems interact. This is why the public policy process for the social consequences of climate change should be reinterpreted through the lens of CT.

CT as a nonlinear approach in physics offers more novel using for realistic perspective to social systems in the last two decades. Public policy and CT have been studied public policy process as complex adaptive system with properties of CT. In light of this, throughout this chapter, public policy process is considered to be a complex adaptive system and three properties of CT are discussed, which are multi-actor, self-organization, and coevolution within steps of public policy.

Depending on the analysis, in order to take an inclusive approach to this dynamic public policy process, governments should embrace constructive components that center the varied goals of multi-actors. In other words, while governments take a flexible approach to agenda setting, they encourage players to develop efficient and effective policy formulations that help to address the social repercussions of climate change. Finally, policy development includes both externalization and internalization within political discourse, which may lead to significant shifts in policy agendas, particularly in climate change.

Furthermore, self-organization properties of CT give fresh perspective to implement policy agenda. Self-organization naturally creates negative and positive feedback systems for continuation or changing of selected policies in a specific issue in social consequences of climate change. Lastly, monitoring and evaluation process is generally fragile in public policy for climate change, which is why CT suggests that coevolution could be key to revisit understanding of the last stage. Coevolution states that systemic monitoring could be useful to hinder connection between diverse policy sectors and more flexible actors such as civil society could be used as facilitator role. Consequently, while individual states and various stakeholders started to lose their willingness on climate change, CT brings new and more realistic perspectives that could foster an innovative way to public policymaking.

## References

Abbass, K., Qasim, M. Z., Song, H., Murshed, M., Mahmood, H., & Younis, I. (2022). A review of the global climate change impacts, adaptation, and sustainable mitigation measures. *Environmental Science and Pollution Research*, *29*(28), 42539–42559.

Açıkalın, Ş. N. (2022). Rethinking actor and order with complexity theory: A novel approach to diplomacy. *Perceptions: Journal of International Affairs*, *27*(2), 177–196.

Beg, N., Morlot, J. C., Davidson, O., Afrane-Okesse, Y., Tyani, L., Denton, F., & Rahman, A. A. (2002). Linkages between climate change and sustainable development. *Climate Policy*, *2*(2–3), 129–144.

Bhattacharya, A. (2020). Rebooting the climate agenda: What should the priorities be? In *Reimagining the global economy: Building back better in a post-Covid-19 world* (pp. 33–37). Global Economy and Development.

Bogg, J., & Geyer, R. (2007). Complexity, science and society. In J. Bogg & R. Geyer (Eds.), *Complexity, science and society*. Radcliffe Publishing Ltd.

Booth, L., Schueller, L. A., Scolobig, A., & Marx, S. (2020). Stakeholder solutions for building interdisciplinary and international synergies between climate change adaptation and disaster risk reduction. *International Journal of Disaster Risk Reduction*, *46*, 101616.

Braunerhjelm, P., & Hepburn, C. (2023). Climate change, complexity, and policy design. *Oxford Review of Economic Policy*, *39*(4), 667–679.

Byrne, D. (2002). *Complexity theory and the social sciences: An introduction*. Routledge.

Cairney, P., & Geyer, R. (Eds.). (2015). *Handbook on complexity and public policy*. Edward Elgar.

Capra, F. (2005). Complexity and life. *Theory, Culture & Society*, *22*(5), 33–44.

Carmona, P., Stef, N., Jabeur, S. B., & Zaied, Y. B. (2023). Climate change and government policy: Fresh insights from complexity theory. *Journal of Environmental Management*, *338*, 117831.

Cimato, F., & Mullan, M. (2010). *Adapting to climate change: Analysing the role of government*. Defra Evidence and Analysis Series. Paper, 1.

Clar, C., Prutsch, A., & Steurer, R. (2013). Barriers and guidelines for public policies on climate change adaptation: A missed opportunity of scientific knowledge-brokerage. *Natural Resources Forum*, *37*(1), 1–18. https://doi.org/10.1111/1477-8947.12013

Da Silva, J., & Lammersen, F. (2017). The aid community should stop pretending to know the answers and start asking the right questions. *OECD insights*. http://oecdinsights.org/2017/01/20/aidand-complexity/

Douglas, E. J., Shepherd, D. A., & Prentice, C. (2020). Using fuzzy-set qualitative comparative analysis for a finer-grained understanding of entrepreneurship. *Journal of Business Venturing*, *35*(1), 105970.

Dunn, W. N. (2017). *Public policy analysis*. Routledge.

Eliadis, F. P., Hill, M. M., & Howlett, M. (Eds.). (2005). *Designing government: From instruments to governance*. McGill-Queen's Press-MQUP.

Erkoçak, E., & Açıkalın, Ş. N. (2015). Complexity theory in public administration and metagovernance. In *Chaos, complexity and leadership 2013* (pp. 73–84). Springer International Publishing.

Friman, M., & Hjerpe, M. (2015). Agreement, significance, and understandings of historical responsibility in climate change negotiations. *Climate Policy*, *15*(3), 302–320.

Gough, I., Meadowcroft, J., Dryzek, J., Gerhards, J., Lengfeld, H., Markandya, A., & Ortiz, R. (2008). Climate change and social policy. *Journal of European Social Policy*, *18*(4), 325–344.

Haynes, P. (2008). Complexity theory and evaluation in public management: A qualitative systems approach. *Public Management Review*, *10*(3), 401–419. https://ssir.org/articles/entry/climate_change_what_is_our_goal

IPCC. (2007). *AR4 Climate change 2007: Synthesis report — IPCC*. IPCC. https://www.ipcc.ch/report/ar4/syr/

IPCC. (2023). Summary for policymakers. In H., Lee & J., Romero (Eds.), *Climate Change 2023: Synthesis Report. Contribution of Working Groups I, II and III to the Sixth Assessment Report of the Intergovernmental Panel on Climate Change [Core Writing Team]* (pp. 1–34). IPCC. https://doi.org/10.59327/IPCC/AR6-9789291691647.001

IPCC. (2024). *History of IPCC*. https://www.ipcc.ch/about/history/#:~:text=The%20Intergovernmental%20Panel%20on%20Climate%20Change%20(IPCC)%20was%20established%20by,UN%20General%20Assembly%20in%201988

Istiadji, A. D., Hardiman, G., & Satwiko, P. (2018). The theory of complexity: A new opportunity to optimize green energy approaches for achieving sustainable development. In *Proceedings of the 18th International Conference on Sustainable Environment and Architecture (SENVAR 2018)* (Vol. 156, pp. 95–105). Atlantis Press.

Jaffe, Y., Bar-Oz, G., & Ellenblum, R. (2019). Improving integration in societal consequences to climate change. *Proceedings of the National Academy of Sciences, 116*(11), 4755–4756.

Jann, W., & Wegrich, K. (2003). Phasenmodelle und politikprozesse: Der policy cycle. In K. Schubert & N. C. Bandelow (Eds.), *Lehrbuch der Politikfeldanalyse* (pp. 71–104). R. Oldenbourg Verlag.

Kastens, K. (2010). *"Going negative on" Negative feedback.* Earth and Mind blog. http://serc.carleton.edu/earthandmind/posts/negativefeedbac.html. Accessed on February 17, 2024.

Kingdon, J. W. (1995). *Agendas, alternatives, and public policies* (2nd ed.). Haper-Collins College Publisher.

Klijn, E. H. (2008). Complexity theory and public administration: What's new? Key concepts in complexity theory compared to their counterparts in public administration research. *Public Management Review, 10*(3), 299–317.

Laukkonen, J., Blanco, P. K., Lenhart, J., Keiner, M., Cavric, B., & Kinuthia-Njenga, C. (2009). Combining climate change adaptation and mitigation measures at the local level. *Habitat International, 33*(3), 287–292.

Lindblom, C. E. (1959). The science of muddling through. *Public Administration, 19*, 79–88.

Luers, A. L., & Sklar, L. S. (2013). The difficult, the dangerous, and the catastrophic: Managing the spectrum of climate risks. *Earth's Future, 2*. https://doi.org/10.1002/2013EF000192

Mach, K. J., Jagannathan, K., Shi, L., Turek-Hankins, L. L., Arnold, J. R., Brelsford, C., Flores, A. N., Gao, J., Martín, C. E., McCollum, D. L., Moss, R., Niemann, J., Rashleigh, B., & Reed, P. M. (2024). Research to confront climate change complexity: Intersectionality, integration, and innovative governance. *Earth's Future, 12*(6). https://doi.org/10.1029/2023ef004392

Macintosh, A., & Wilkinson, D. (2016). Complexity theory and the constraints on environmental policymaking. *Journal of Environmental Law, 28*(1), 65–93.

Mason, R. B. (2007). The external environment's effect on management and strategy: A complexity theory approach. *Management Decision, 45*(1), 10–28.

McMillan, E. (2003). *Complexity, organizations and change* (Vol. 1). Routledge.

Meek, J. W. (2010). Complexity theory for public administration and policy. *Emergence: Complexity and Organization, 12*(1), 1.

Memiş O. B., & Aydin, R. (2023). İklim deäiş ikliäinin ekonomik büyüme üzerine etkisi: Türkiye örneäi. *Third Sector Social Economic Review, 58*(2), 1065–1080.

Mitleton-Kelly, E. (2000). Complexity: Partial support for BPR? In *Systems engineering for business process change: Collected papers from the EPSRC research programme* (pp. 24–37). Springer.

Mitleton-Kelly, E. (2003). *Complex systems and evolutionary perspectives on organizations: The application of complexity theory to organizations.* Elsevier Science Ltd.

Morçöl, G. (2010). Issues in reconceptualizing public policy from the perspective of complexity theory. *Emergence: Complexity and Organization, 12*(1), 52.

Morçöl, G. (2012). *A complexity theory for public policy.* Routledge.

Morçöl, G. (2013). *A complexity theory for public policy.* Routledge.

Mukherjee, V., & Mustafa, F. (2019). Climate change and right to development. *Management and Economics Research Journal, 5.*

Pongiglione, F., & Cherlet, J. (2015). The social and behavioral dimensions of climate change: Fundamental but disregarded? *Journal for General Philosophy of Science*, *46*, 383–391.

Safonov, G. (2019). *Social consequences of climate change: Building climate friendly and resilient communities via transition from planned to market economies: Climate change, energy and environment*. Friedrich-Ebert-Stiftung, Department for Central and Eastern Europe.

Tomé, L., & Açıkalın, Ş.N. (2019). Complexity theory as a new lens in IR: System and change. In Ş. Erçetin & N. Potas (Eds.), *Chaos, complexity and leadership 2017. ICCLS 2017. Springer Proceedings in Complexity*. Springer. https://doi.org/10.1007/978-3-319-89875-9_1

Urry, J. (2003). *Global complexity*. Polity.

Chapter 11

# The Power of Education in Tackling Climate Change: Thoughts for Future

*Duygu Sönmez*

Hacettepe University, Turkey

## Abstract

Climate change is an issue which has been in the global agenda since late 19th century. Although it is not a new issue, its urgency requires every nation, agency, and individual to take actions; otherwise, the consequences would be dire. In that respect, a global effort is in place to tackle the climate change with protocols, agencies, and educational reforms. Education is considered one of the greatest powers we possess against climate change. Climate change is a complex topic, and using systems understanding is recommended in climate change education and education for sustainable development. Climate change education aims to develop an understanding, increase knowledge and skills among individuals so that we can acquire a pro-environmental behavior.

*Keywords*: Climate change; education for sustainable development; climate change education; sustainability; science education

## Introduction

Climate change is one of the topics we have been discussing since late 19th century. Starting with concerns on the increasing levels of carbon dioxide ($CO_2$) in 1800s and continuing with today's concerns of passing the tipping point and how to build a sustainable future, climate change has a long history. Before discussing climate change further, we need to define what climate change is. National Aeronautics and Space Administration (NASA) defines climate change as "a long-term change in the average weather patterns that have come to define Earth's local, regional and global climates. These changes have a broad range of observed effects that are synonymous with the term" (NASA, 2024, para. 1).

**The Social Consequences of Climate Change, 127–139**

**Published under exclusive licence by Emerald Publishing Limited**
**doi:10.1108/978-1-83797-677-520241011**

The contributors to climate change are mainly human-driven activities, and NASA defines such activities as use of fossil fuels which has raised surface temperature and human activities that overwhelms natural processes. Since 1900s, the estimated temperature increase on Earth, due to human activities, is about 1°C in comparison to preindustrial period (NASA, 2024).

We are living in a self-sustaining biosphere; however, most individuals do neither recognize the finite resources Earth has been providing us nor realize that overwhelming human activities cause Earth systems to collapse without getting enough time and opportunity to recover. Tipping point is the terminology we use to describe that stress and earth's inability to recover. In layman's terms, Intergovernmental Panel on Climate Change (IPCC) defines tipping point as "critical thresholds in a system that, when exceeded, can lead to a significant change in the state of the system, often with an understanding that the change is irreversible" (IPCC, 2014, para. 1). Lenton et al. (2019) state that the initial idea of 5°C or more increase in global temperatures would lead to exceeding of tipping points, at the time the idea of tipping point first introduced by the IPCC, is now re-evaluated, and it is believed that even between 1 and 2°C of increase would cause the tipping points to happen. Loss of amazon rainforest, ice collapse, biosphere boundaries such as loss of corals are some of the tipping points experts warn us about (Lenton et al., 2019). Experts argue on the threats that are evidence to exceeding of tipping points and call for an emergency response at all levels. One of their major focus points is the knowledge gaps on the issue. How can we overcome knowledge gaps among individuals and communities? What kind of actions do we need to take?

## A Brief Look into Climate Change Policies

Climate change is a global issue; therefore, the solution has to come from a global perspective as well. But in the process, individual and community-based actions are needed as well nationwide and international responses. In the history of climate change, there have been many attempts to tackle the issue through protocols and organizational bodies reviewing and providing reports and suggestions on the matter to support national and international efforts.

In 1988, the IPCC was established by the United Nations as a part of Environment Programme (UNEP). Initially, IPCC was established to review "the state of knowledge of the science of climate change; the social and economic impact of climate change" and provide recommendations and strategies accordingly (IPCC, 2024, para. 5).

In 1992, United Nations Framework Convention on Climate Change (UNFCCC) was adapted at the Earth Summit in Rio de Janeiro. UNFCCC provides a framework for international cooperation on climate change. The most recent United Nations Climate Change Conference (COP28) took place in December 2023. The focus was how to speed up the progress in regards to climate action which has been too slow until now. Kyoto protocol was adapted in 1997; currently, number of parties who signed the protocol is 192. Through this protocol, industrialized countries agree to lower their greenhouse gases (GHGs) emissions to a certain level. Kyoto protocol is based on the principles of UNFCCC (United Nations Climate Change, 2024).

United Nations Decade of Education for Sustainable Development (UN DESD) – 2005–2014 – aims to promote sustainable development in education. This resolution is adapted following Johannesburg Summit which took place in 2002. The main goal is to achieve sustainable behavior through integration of content and values of sustainable development in education (Liimatainen, 2013). DESD's vision is as follows "where everyone has the opportunity to benefit from education and learn the values, behavior and lifestyles required for a sustainable future and for positive societal transformation" (UNESCO, 2005, p. 6). Aligned with their vision, DESD focuses their work on program development and inclusion of Education for Sustainable Development through collaboration with governments as well as focusing on teacher education (Mulà & Tilbury, 2009).

In 2009, UNESCO Climate Change Initiative was adopted in Copenhagen during the 15th Conference of the Parties to the UNFCCC (UNFCCC COP15). UNESCO Climate Change Initiative has four thematic areas: educational, scientific, ethical, and environmental, and aims for increasing mitigation and adaption capacities of countries through an interdisciplinary collaboration (UNESCO, 2010). Core program objectives of climate change education by UNESCO (2010) are as follows:

- "Strengthen the capacity of Member States to provide quality climate change education for sustainable development at primary and secondary school level through improved education policy, analysis, research and planning, teacher education and training of education planners, training on curriculum review/ reform."
- "Encourage and enhance innovative teaching approaches to integrate quality climate change education for sustainable development in school through interdisciplinary practices, science education, whole school approaches, technical and vocational education and training (TVET), education on disaster risk reduction."
- "Raise awareness about climate change and enhance non-formal education programmes through media, networking and partnerships" (p. 5).

UNESCO has over 30 programs aiming to contribute to scientific knowledge with the aim of educating present and future generations about climate change and implications of our actions from an ethical perspective, and national and global efforts are taking place to resolve the climate change issue (UNESCO, 2021).

The Paris Agreement (2015) was adapted in 2015 for the purpose of lowering "the increase in the global average temperature to well below 2°C above pre-industrial levels" (para. 2) and pursue efforts "to limit the temperature increase to 1.5°C above pre-industrial levels" (para 2). The Paris Agreement is an international treaty and has a binding nature for the 196 parties. Countries support each other during this long-term endeavor through financial assistance, technology exchange, and support in capacity building for mitigation and adaptation (UNCC, 2024).

Legislations and flagships laws are how countries take actions legally against climate change and responses are changing and evolving as the climate change is becoming a pressing issue. European Union has adapted 25 climate change-related legislations since 2013, while in Jamaica and Nepal only three. As seen from these two examples progress takes different approaches and pace among countries (Townshend et al., 2013). Townshend et al. (2013) point out that number of legislations or flagship laws cannot necessarily state a country's response to climate change since countries can use regulatory approaches to deal with the issue. Legislations and regulatory action have an impact on governmental authorities as well as other sectors such as energy and industry including focusing on management and preservation (Townshend et al., 2013). Any effort to reduce the concentration of GHGs is a valid and necessary effort. However, these acts of mitigations are seemed not enough, and we need to modify our social, economic, and ecological systems to prevent vulnerability of the ecosystem (Boon, 2016). The term "green" found its place in terms of these attempts of adaptation meaning not just targeting for a lower carbon emission in current state but also aiming for a structural change that is low on carbon (Townshend et al., 2013). Scientific understanding of climate change is vital but, as stated by UNESCO (2010), future will be shaped by the children of today. Developing an awareness and understanding on climate change among citizens and specifically children would be one of the best strategies to change attitudes and sustainable development.

A public awareness against climate change started to develop as of 2000s that is powerful as it can support and act as a driving force for action for schools, communities, and even governments (Stevenson et al., 2017). In 2018, we witnessed a new movement against climate change initiated by youth, Youth Climate Movement (YCM). During the movement, names such as Greta Thunberg became the face of the movement on media. An initiative started from schools is now evolving along with the geopolitical changes in the world such as COVID-19 and the war in Ukraine to keep the momentum they have created up. Other movements and organizations are also being established such as Climate Vanguards which is a youth-led think tank (Herfort et al., 2023), attempts to make a difference.

## Climate Change Education

Humans have never been so distant to the earth we live in, in the human history as we are today. Consequently, human actions of industrialized age have been quite damaging on Earth, and the future has never been this unknown. We do not know if our children, grandchildren, and grand grandchildren will have the opportunity to experience the air, the land, and the seas as we have experienced so far. Boon (2016) argues that future generations need to develop necessary skills so that they can be prepared to the sustainability issues such as loss of biodiversity, proliferating number of human populations, scarcity of resources, and climate change.

We urgently need to face the elephant, climate change, in the room and take necessary actions since it cannot be ignored anymore.

Climate change is a complex topic with an interdisciplinary nature. Therefore, climate change education should be using systems understanding addressing all scientific disciplines such as ecological, social, political, and so on (Mochizuki & Bryan, 2015). In addition to the complex nature of climate change, there are various sources of information that requires ability to use cognitive skills which can be achieved through education. The digital era we are living in makes it possible to access a vast amount of information, true–false or old and current. Since what we know and accept as truth determines our actions, an accurate understanding of climate change is vital. And in a bigger scale, our knowledge, understanding, and behaviors are reflected on the societies. Therefore, every individual should be able to reflect on and evaluate conflicting information (Bråten et al., 2011), decide on the accuracy, and use it accordingly which necessitates climate literacy. These changes at the individual level are also expected to be reflected on institutions and governments as a ripple effect in the long run as a result of climate change education (Mochizuki & Bryan, 2015).

Climate change education aims to develop an understanding and increase knowledge and skills among individuals so that we can acquire a behavioral change, first at the individual level and then at the societal level. One of the focuses of climate change education is sustainable development. And a sustainable life style requires a significant shift in our mindset which cannot be achieved without education. Formal and informal education have a big role to play in our understanding and interaction with environment in a conscious manner as (i) it helps us to develop necessary skills and knowledge to understand environment, (ii) develop environment friendly behavior and life style, (iii) come up with solutions to environmental problems for a sustainable lifestyle, and (iv) future and learning traditional knowledge from local communities which can help us to adapt climate change and develop an intercultural respect (UNESCO, 2016). On the other hand, from a formal perspective, climate change education focuses on a specific issue and aims for understanding climate change, the risks associated with it, how our actions contribute to the problem, and encourages us to ask critical questions and develop correct responses and actions to the problem (Stevenson et al., 2017). A climate literate person understands the "climate's influence" on them and the society and their "influence on climate" (U.S. Global Change Research Program, 2009).

Studies conducted to investigate our understanding about climate change portray an ill-understood concept regardless the age group or geographical location. Research on how climate change and related issues are understood and what kind of attitudes are held against climate change targets two main groups, students (K-16) and teachers (preservice and in-service). Boon (2010) states that complexity of science behind climate change is one of the reasons for misunderstanding and misconceptions. A solid education at all grade levels (K-16) is required for a comprehensive understanding of climate change and related concepts. Teachers are a particular interest to researchers since they are ones either teaching or will teach climate change in their classrooms, and it's important that they have an accurate and scientific understanding of the topic. Large-scale comparative research, reports, and studies investigating trends portray us a picture about climate change education.

Climate education research between the years 1991 and 2018 shows a changing trend of interest. The distribution of studies investigating climate literacy and education between these years were 2010–2018 (71.7%), 2001–2009 (21.7%), and 1991–2000 (6.5%) (Bhattacharya et al., 2021). As seen from the results, there is a significant emphasis on climate change in research since 2010. Between these years, out of 178 research from 27 countries, 128 of them focus on K-16 grade level and 50 of them on teachers (Bhattacharya et al., 2021). Most of the research on climate change education takes place in the United States (US) followed by Canada, Europe, Australia, Africa, and although significantly less in numbers by Asia and South America (Rousell & Cutter-Mackenzie-Knowles, 2020). These researches investigate students' and teachers' conceptions about and understanding of global climate change, greenhouse effect, global warming, and ozone layer as well as the relationship between scientific knowledge and pro-environmental behavior. Other topics included in research were educators' beliefs, their readiness to teach the content, curricula, and climate education programs. Until the 2000, the general interest was to investigate students' conceptions of climate-related concepts such as global warming in terms of science education or climate change education. But starting from 2000s, the interest shifted to learning aspect of climate change education, specifically how students learn and conceptualize climate change and what are the effective factors (Bhattacharya et al., 2021; Rousell & Cutter-Mackenzie-Knowles, 2020). Research interests also exhibit geographic variation. In the United States there is a focus on climate change curricula and indigenous knowledge, while Canadian and European focus was towards pedagogical and social initiatives within formal education and informal institutions such as museums. Australia and Africa similarly emphasize climate change education curricula, whereas Pacific Islands focus concentrate on community-based climate change education (Rousell & Cutter-Mackenzie-Knowles, 2020).

The general assumption between scientific knowledge and pro-environment behavior is that they are positively correlated. But as Dijkstra and Goedhart (2012) state, there is a complex relationship between them which is difficult to achieve and prove through empirical studies. Which approach or way of teaching effects behavior? Answer to this problem is quite an important one as our ultimate goal is to change attitudes and behavior toward climate change through education. Devine-Wright et al. (2004) consider the issue from a psychological perspective and investigated role of self-efficacy and role of situational influences in terms of pro-environmental behavior in particular to the energy issues. They report positive influence of cooperative learning environments. Positive impact of other learning environments such as socio-constructivist and experiential learning (Pruneau et al., 2003) and constructivist approach encouraging learning and critical thinking (Bardsley & Bardsley, 2007) are also reported in literature regarding climate change education. Identifying effective strategies that are positive on the development of pro-environmental behavior is important for our

NASA. (2024, February 2). *What is climate change?* https://climate.nasa.gov/what-is-climate-change/#:~:text=Facts%E2%80%BAWhat%20Is%20Climate%20Change, are%20synonymous%20with%20the%20term

National Research Council. (2012). *A framework for K-12 science education: Practices, crosscutting concepts, and core ideas*. National Academies Press.

Next Generation Science Standards Lead States (NGSS). (2013). *Next generation science standards: For states, by states*. National Academies Press.

Pallant, A., & Lee, H. S. (2015). Constructing scientific arguments using evidence from dynamic computational climate models. *Journal of Science Education and Technology*, *24*, 378–395.

Pruneau, D., Gravel, H., Bourque, W., & Langis, J. (2003). Experimentation with a socio-constructivist process for climate change education. *Environmental Education Research*, *9*(4), 429–446.

Rousell, D., & Cutter-Mackenzie-Knowles, A. (2020). A systematic review of climate change education: Giving children and young people a 'voice' and a 'hand' in redressing climate change. *Children's Geographies*, *18*(2), 191–208.

Stevenson, R. B., Nicholls, J., & Whitehouse, H. (2017). What is climate change education? *Curriculum Perspectives*, *37*, 67–71.

The Paris Agreement. (2015, February 12). *United Nations Climate Change*. https://unfccc.int/process-and-meetings/the-paris-agreement

Townshend, T., Fankhauser, S., Aybar, R., Collins, M., Landesman, T., Nachmany, M., & Pavese, C. (2013). How national legislation can help to solve climate change. *Nature Climate Change*, *3*(5), 430–432.

Turhan, E., Cerit Mazlum, S., Şahin, Ü., Şorman, A. H., & Cem Gündoğan, A. (2016). Beyond special circumstances: Climate change policy in Turkey 1992–2015. *Wiley Interdisciplinary Reviews: Climate Change*, *7*(3), 448–460.

Türkiye Cumhuriyeti Cumhurbaşkanlığı Strateji ve Bütçe Daire Başkanlığı (SBB). (2023). *Onbirinci kalkınma planı (2019–2023)*. https://www.sbb.gov.tr/wp-content/uploads/2022/07/On_Birinci_Kalkinma_Plani-2019-2023.pdf

UNESCO. (2010). *UNESCO climate change initiative: Climate change education for sustainable development*. Author.

UNESCO. (2015). *Not just hot air: Putting climate change education into practice*. Author.

UNESCO. (2016). *Education for people and planet: Creating sustainable futures for all*. Author.

UNESCO. (2021). *Learn for our planet: A global review of how environmental issues are integrated in education*. Author.

United Nations Educational, Scientific and Cultural Organisation (UNESCO). (2005). *United Nations decade of education for sustainable development (2005–14): International Implementation Scheme*. UNESCO.

U.S. Global Change Research Program. (2009). *Climate literacy: The essential principles of climate science*. US Global Change Research Program. https://downloads.globalchange.gov/Literacy/climate_literacy_highres_english.pdf

United Nations Climate Change. (2024, February 12). *What is the Kyoto Protocol?* https://unfccc.int/kyoto_protocol#:~:text=In%20short%2C%20the%20Kyoto%20Protocol,accordance%20with%20agreed%20individual%20targets

Chapter 12

# Analysis of Global Development Goals in the Context of Istanbul

*Metin Erol*[a] *and Burak Kaplan*[b]

[a]Istanbul University, Türkiye
[b]Marmara University, Türkiye

## Abstract

These are the Municipality Law No. 5216 and the Metropolitan Municipality Law No. 5393. The legislative body has determined it as a requirement for metropolitan municipalities to prepare strategic plans through the Public Financial Management and Control Law No. 5018. In this study, the compliance of the strategic plans prepared by Istanbul Metropolitan Municipality (IMM) for the period 2015–2019 and 2020–2024 with the 17 Global Development Goals of the United Nations will be examined. As a result of the study, it was aimed to compare the compatibility of both plans, which were prepared with different political understandings, with the 17 Global Development Goals. As a method in the study, the aims and objectives of the strategic plans of IMM in different periods will be codified and then their compliance with the 17 Global Development Goals will be checked. Ultimately, the study aims to contribute to the public policy literature on the responsibility of public institutions on a scale ranging from local to global in achieving policy goals.

*Keywords*: Climate change; environment; metropolitan municipality; strategic plan; sustainability

## Introduction

The United Nations (UN), which was established with the charter signed on June 26, 1945, after the Second World War, with the aim of making the international system peaceful and reliable, strengthening the global justice mechanism, and improving cooperation among countries in social, economic, and humanitarian

**The Social Consequences of Climate Change, 141–161**

**doi:10.1108/978-1-83797-677-520241012**

problems, has 193 members as of 2024 (United Nations, 2024a). Since its establishment, the UN has undertaken a number of initiatives to solve common problems of humanity. To give an example, the UN laid the foundations of modern environmental law by addressing the relationship between humans and the environment with the Stockholm Conference held in 1972, 27 years after its establishment. In the Stockholm Declaration published as a result of the conference, the main emphasis was on sensitivity toward protecting the environment (Sand, 2007, pp. 33–35). The UN's work on the common problems of humanity continued with the Human Settlements Program (HABITAT) summits. The program, which was initiated to provide adequate housing for everyone in residential areas, is organized every 20 years. These summits, the first of which was held in Vancouver in 1976, the second in Istanbul in 1996, and the third in Kito in 2016, focused on the right to housing and the questions and problems arising from the urbanization trend (United Nations, 2024b). Another of the UN's studies on environmental sensitivity is the report titled Our Common Future prepared by the UN World Commission on Environment and Development. In the report, for the first time, the concept of sustainable development was defined as "humanity has the ability to make development sustainable to ensure that it meets the needs of the present without compromising the ability of future generations to meet their own needs" and was examined in its economic, social, and environmental dimensions (United Nations, 1987).

The sensitivity of the UN toward the future of the world also showed itself with the conferences held in Rio de Janeiro, the capital of Brazil. With the Rio, Rio +5, Rio +10, and Rio +20 conferences, emphasis was placed on environmental protection and the inclusion of sustainability principles in all management processes. In the first series of these conferences, held in 1992, the United Nations Framework Convention on Climate Change was opened for signature and entered into force on March 21, 1994, with the confirmation of the countries. It has been observed that this agreement, which demonstrates the goodwill of the party countries to combat climate change, began to concrete its goals with the Kyoto Protocol signed in 1997 (United Nations, 2024c, 2024d).

One of the recent examples of the UN's efforts to protect the future of humanity and the environment is the "2030 Sustainable Development Agenda" adopted by the United Nations member states in 2015. Within the scope of this agenda, a plan based on the peace and prosperity of people and the planet today and in the future has been put forward. The 17 global goals determined within the plan are "*No Poverty, Zero Hunger, Good Health and Well-Being, Quality Education, Gender Equality, Clean Water and Sanitation, Affordable and Clean Energy, Decent Work and Economic Growth, Industry, Innovation and Infrastructure, Reduced Inequalities, Sustainable Cities and Communities, Responsible Production and Consumption, Climate Action, Life Below Water, Life on Land, Peace, Justice, and Strong Institutions, Partnerships for the Goals*". Various studies have been implemented in line with a total of 169 targets determined for each of these objectives (United Nations, 2015b).

At the heart of these studies carried out by the UN are the risks that the planet will face in the future. As a matter of fact, the rapid increase in the world population and the consumption pressure on natural resources are threatening the future of humanity. On the other hand, the dynamics of industrialization and urbanization also turn into an output in the form of air, water, soil, and noise pollution. The ever-increasing consumption of fossil fuels, the gradual decrease in forest areas, and the accumulation of greenhouse gases in the atmosphere that cause global climate change are the results of today's world's consumption and lifestyle. This process that the world is experiencing directly affects weather events. Sudden changes in air temperature, rising average air temperature values, or sudden and intense precipitation that can turn into disasters are undoubtedly projections of the cycle experienced in the ecosystem. Scientists explain this change with global temperature changes. In this context, it is calculated that the global temperature value will increase between 1 and 3.5 C° and the average sea level will rise between 15 and 95 cm by 2,100.

It is important that this process that the world is experiencing and will continue to experience is managed with a public policy set based on cooperation at international and national scales. Supporting these efforts led by the UN with policies to be implemented at national and local levels is a must to achieve global success. As a matter of fact, it is important to support global goals with projects or policies implemented, especially starting from the local level. In this context, one of the most basic indicators is the increasing urban population in the world. While the global urbanization level was 3% in the 1800s, this level reached up to 47% in the 21st century. When population projections are examined, it is calculated that two out of every three people will live in cities by 2030 (United Nations, 2015d). This rapid increase in urbanization has the potential to further determine the impact of cities, which account for 80% of greenhouse gas emissions, on global climate change in the future.

There is two-way relationship between global climate change and urbanization. While the activities carried out in cities negatively affect global climate change, it is the urban system that is most affected by the negative consequences of global climate change. Therefore, city managers should consider the effects of the decisions on the city when deciding on the policies they will implement with a long-term perspective. The long-term planning and management understanding of cities against the consequences of global climate change is only possible with a strategic management approach. In this respect, municipalities determining their goals and objectives and implementing their services with a strategic management approach means that the city is seen as a trust to be left to future generations.

The long-term planning and management approach of cities against the consequences of global climate change is only possible with a strategic management approach. In this respect, the fact that the municipalities, which are primarily responsible for the provision of urban services, determine their goals and objectives and implement their services and projects with a strategic management approach means that the city is seen as a trust to be handed over to future generations.

The institutionalization of the strategic management approach on behalf of municipalities in Türkiye took place with the Public Financial Management and Control Law No. 5018. In the law published in the Official Gazette on December 24,

2003, the "strategic plan" is defined as "*the plan that includes the medium and long-term goals of public administrations, basic principles and policies, targets and priorities, performance criteria, methods to be followed to achieve these and resource allocation*" (Official Gazette of the Republic of Türkiye, 2003). In addition, in the Metropolitan Municipality Law No. 5216, the preparation of a strategic plan is among the duties of the metropolitan municipality.

In this study, the effects of the 17 Global Development Goals will be discussed, and the compatibility of these goals with the strategic plans prepared by Istanbul Metropolitan Municipality (IMM) for the periods 2015–2019 and 2020–2024 will be analyzed. As a result of the study, it is aimed to compare the plans prepared in these periods when IMM was governed under different political understandings. Ultimately, it is expected that the study will contribute to the public policy literature on the impact of policymaking processes.

This study consists of three parts. In the first part, 17 Sustainable Development Goals (SDGs) were analyzed in the context of the global climate crisis. It was focused on what kind of topics these targets adopted by the UN, which aim at sustainable development on a global scale until 2030, touch upon in the fight against global climate change. In the second part, the importance of national policymakers in combating global climate change was explained through strategic plans. In this regard, the development of the strategic plan understanding in Türkiye has been revealed in the context of its theoretical background and legal regulations. Finally, in the third chapter, the strategic plans of IMM, which were determined as the case studies of the study, were analyzed in the context of the global climate crisis, and the strategic plans of the 2015–2019 and 2019–2024 periods were compared.

## Sustainable Development Goals Against the Global Climate Crisis

With the decision taken by all delegates at the UN General Assembly in September 2000, Millennium Development Goals were determined to improve social, economic, and environmental conditions, combat poverty globally, and promote human dignity. All participants agreed on eight targets that should be developed by 2015 with the Millennium Development Goals. These goals are (United Nations, 2015a):

- Eradicating extreme poverty and hunger.
- Universal access to basic education.
- Reducing gender inequality.
- Reducing child deaths.
- Improving maternal health.
- Fighting diseases such as HIV/AIDS and malaria.
- Ensuring environmental sustainability.
- Developing a global partnership for development.

Following the monitoring and evaluation of the Millennium Development Goals at the Sustainable Development Summit held by the UN in 2015, it was concluded that these targets could not provide solutions to the economic, social, and environmental problems faced by the world. Therefore, a consensus has been reached that the SDGs should be comprehensive and capable of providing solutions to the world's problems (Peşkiricioğlu, 2016, p. 6).

193 UN member countries, including Türkiye, have adopted the 2030 SDGs. Thus, for sustainable development until 2030, 17 Goals and 169 Sub-Targets were determined under various headings such as basic human needs, climate change, environmental protection and cooperation between institutions. In line with these goals, it is envisaged that countries will comply with these goals in their development plans, programs, and strategic objectives. The 17 SDGs determined by the UN until 2030 are as seen in Fig. 12.1 (United Nations, 2015b).

The features of the new agenda that emerged as a result of this change from Millennium Development Goals to SDGs are as follows (Kavas, 2016):

- With the new agenda, the concept of sustainability has been placed at the center of development.
- The concept of climate change has been replaced by the concept of green growth.
- A universal development approach has been adopted, abandoning the definition of rich, poor, or middle-income countries.
- In addition to cooperation between states and nongovernmental organizations, the business, technological, and scientific worlds have also been involved.

Although global climate change is not an issue directly mentioned with the new agenda, the SDGs are directly or indirectly linked to global climate change. Some of the new targets determined deal with the factors that lead to global climate change, while the other part deals with the consequences of global climate change. At the same time, it is seen that some of these goals are two-dimensional, that is, they both contribute to the fight against global climate change and combat the consequences caused by global climate change (United Nations, 2015c).

| | | | | | |
|---|---|---|---|---|---|
| *No Poverty* | *Zero Hunger* | *Good Health And Well-Being* | *Quality Education* | *Gender Equality* | *Clean Water And Sanitation* |
| *Affordable and Clean Energy* | *Decent Work and Economic Growth* | *Industry, Innovation and Infrastructure* | *Reduced Inequalities* | *Sustainable Cities and Communities* | *Responsible Production and Consumption* |
| *Climate Action* | *Life Below Water* | *Life on Land* | *Peace, Justice and Strong Institutions* | *Partnerships for the Goals* | *SUSTAINABLE DEVELOPMENT GOALS* |

Fig. 12.1. Sustainable Development Goals.

For example, the first of the SDGs, No Poverty, points out the fight against conditions that arise as a direct result of global climate change. If the current trend in the world continues, it is estimated that 575 million people will live in extreme poverty in 2030. In order to combat these conditions, it seems that the subgoals determined are aimed at preventing income inequality in the world. Even though this situation does not seem to be directly related to global climate change, it should be accepted as an outcome of the colonial order created by the current international system. As a matter of fact, the fact that militarily and politically powerful countries strengthen their economic structures while ignoring countries other than themselves and their allies has led to a significant portion of other countries being underdeveloped. Therefore, even though this goal does not seem to be related to climate change, it can be interpreted as an intention to alleviate the damage to the environment and climate caused by the structural problems caused by the international system (United Nations, 2015c).

Secondly, the Zero Hunger goal focuses on combating the consequences of global climate change, such as the decrease in agricultural land and difficulty in accessing food (United Nations, 2015c). The third Good Health and Well-Being goal is aimed at the negative conditions experienced as a result of global climate change. The impact of climate change on health occurs indirectly. Heat waves, floods, storms, and sudden temperature changes lead to consequences such as an increase in infectious diseases and difficulty in accessing healthy food (Sukkar, 2021; Çelik et al., 2008, p. 10). This situation negatively affects segments with low-income levels, whether they live in urban or rural areas (United Nations, 2015c).

It is not possible to link the fourth goal, Quality Education, with the fifth goal, Gender Equality, with global climate change. It is possible to see these targets as determined as a result of the understanding of sustainable development on a global scale, which is the center of the new agenda (United Nations, 2015c). The sixth goal, Clean Water and Sanitation, is a result of global climate change because the increase in temperature causes the underground water level and wetlands to decrease (Wheeler & Braun, 2013, p. 509; Ayva et al., 2023, p. 48). Especially in agricultural areas, fertilization and pesticide activities implemented to compensate for the negativities caused by this situation threaten the sustainability of water resources by causing pollution of diminished water resources (United Nations, 2015c).

The seventh goal, Affordable and Clean Energy, stands out as a strategically important topic in the fight against global climate change. It has been understood that the increasing use of fossil fuels after the Industrial Revolution has a high cost to the future of the environment and humanity. The increase in the level of carbon accumulated in the atmosphere threatens the ecosystem and leads to global climate change. Therefore, this goal aims to replace traditional energy sources that cause global climate change with green, that is, environmentally friendly energy sources (Gernaat et al., 2021, p. 119; Keskin, 2019). The contribution of increasing the share of renewable energy sources such as wind, sun, and biomass in the energy supply to the protection of the natural environment and the reduction of carbon accumulation in the atmosphere has a significant share in the fight against global climate change (United Nations, 2015c).

The eighth goal is Decent Work and Economic Growth, the ninth is Industry, Innovation and Infrastructure, the 10th is Reduced Inequalities, and the 11th is Sustainable Cities and Communities. Just like the fourth and fifth goals, it is difficult to link them to global climate change. It is possible to accept these targets as a reflection of the global sustainable development approach determined in accordance with the nature of the new agenda (United Nations, 2015c). The 12th goal, Responsible Production and Consumption, aims to change consumption behaviors that lead to global climate change by focusing on the pressure caused by humanity's consumption habits on natural resources (Gunawan et al., 2019, pp. 1–2; Çelik & Küçük, 2020, pp. 11–12). Today's consumption habits have many environmental effects, such as directly or indirectly increasing carbon emissions, reducing biodiversity, causing global warming, and depleting natural resources (United Nations, 2015c).

The 13th goal, Climate Action, is a goal that directly puts global climate change on the agenda. With this goal, it is aimed to create resilient settlement areas all over the world against the climate-related hazards and natural disasters that the world faces. In this regard, importance has been given to integrating global targets into the national policies and strategies of countries in the fight against climate change. For countries to contribute to this goal on a national scale, certain responsibilities have been assigned to developed countries. The most important of these is the 100 billion dollar financing brought to developed countries that are parties to the UNFCCC, which they pledge to provide so that developing countries can successfully complete the green transformation. As a result, this goal refers to national responsibilities in a struggle to be carried out on a global scale and to global cooperation to fulfill national responsibilities (United Nations, 2015c).

The 14th goal, Life Below Water, focuses on the negative impacts of marine life from human activities. In particular, the increase in sea temperatures due to global warming negatively affects marine biodiversity. In this sense, global climate change directly affects seas, fish, habitat ecosystems, fish stocks, fisheries, and aquaculture. For example, as a result of the warming of seawater, negativities occur in the lives of some aquatic species. Atlantic Salmon is known to be the most affected species. Since the melting of glaciers affects summer and autumn currents, this species becomes confused in its migration routes (Hoegh-Guldberg & Bruno, 2018, p. 1523; Kayhan et al., 2019, p. 129). Again, the seawater temperature being above the most suitable temperatures for this species makes this species susceptible to parasites and diseases (United Nations, 2015c).

The 15th goal, Life on Land, has been determined as a goal that prioritizes combating the problems caused by global climate change in the ecosystem. It is desired to protect and ensure sustainable use of forests and wetlands, terrestrial and inland freshwater ecosystems in line with international agreements (Menga, 2023; Demir, 2009, p. 52). As a result, this goal can be seen as an effort to enrich biodiversity damaged as a result of global climate change and to protect the environment in terms of terrestrial life (United Nations, 2015c).

As a result, the 16th goal, Peace, Justice, and Strong Institutions, and the 17th goal, Partnerships for the Goals, are not topics directly or indirectly related to global climate change. Peace, Justice, and Strong Institutions goal aims to build social and lasting peace and ensure that everyone has access to justice institutions. In addition to these objectives, attention has been paid to ensuring that institutions are accountable, inclusive, and effective. The Partnerships for the Goals addresses the need to revitalize global partnerships to bring the SDGs into practice (United Nations, 2015c).

The 17 Global Development Goals set by the UN for sustainable development by 2030 are directly or indirectly related to climate change. It seems that this relationship is two-dimensional. The primary dimension is that Global Development Goals focus on the development of policies to be implemented to address the problems or consequences of climate change; the secondary dimension is that Global Development Goals are aimed at creating policies to prevent the root factors that cause climate change.

Fighting the problems caused by climate change should be considered a global agenda. As a matter of fact, an environmental problem experienced in any region of the world has the potential to turn into a regional or global crisis without knowing national borders. Due to this situation, the need for national solutions and global collaborations is increasing day by day. It is possible to see goals and targets that bear responsibility for solving global problems in national policy documents and even in the strategic plans of cities prepared on a local scale.

Strategic Plans as a Tool of National Challenge Against the Global Climate Crisis.

The concept of strategic planning has gained ground in the public sector as a result of discussions such as the reorganization of relations between central and local administration, democratization, and localization of public administration that emerged through new public management. It is an understanding that indicators such as customer satisfaction and performance-oriented management, which are important in the private sector, should also have their equivalent in the public sector. This new understanding, in which customer satisfaction in the private sector is replaced by citizen satisfaction, has been built as a new culture that aims to manage public institutions in line with the understanding of efficiency and accountability (Favoreu et al., 2015, p. 1; Demirkaya, 2015, pp. 15–29, p. 16).

The planning culture that is tried to be built within public administration has similar characteristics to the assumption that public administration will be more efficient as a result of being organized according to bureaucratic principles. In other words, perceiving as more formality and detailism carries a risk that may result in public institutions becoming cumbersome. With this structure, the output resulting from the expected effect of structuring according to the principle of bureaucracy in public administration, that is, the cumbersome and inefficient structure, is also a threat to the planning approach. In order to prevent this threat from becoming a reality, an understanding in which citizen demands and efficiency are prioritized has been adopted in the new public management (Berry, 1994, p. 323). Therefore, the risks inherent in the planning approach for classical public administration have undergone radical changes with the new public management, aiming to create an effective and efficient public organization (Pollitt & Bouckaert, 2011).

It is possible to see this culture as the desire of institutions to have a more successful structure. However, the success or failure of institutions depends on different results depending on the strategy they have prepared and their compliance with this strategy. It is clear that just as a correctly prepared and successfully implemented strategy can lead institutions to success, incorrectly prepared strategies can also lead to the end of institutions. It is vitally important for public institutions to consider the effectiveness of the plans they will prepare in this sense and their value in terms of performance management. It is possible to summarize the advantages that public institutions will gain through strategic planning as follows (Demirkaya, 2019, p. 11):

- *"Describing the path between the point where the institution is and the situation it wishes to reach*
- *Requiring the institution to determine its goals, targets and the methods that will enable them to be achieved*
- *Having a long-term and future-oriented perspective*
- *Preparing the institution's budget in a way that expresses the goals and targets set out in the strategic plan*
- *Guiding resource allocation based on priorities and accountability*
- *Developing the tendency to collect and analyze data*
- *Providing effectiveness and efficiency in service and resource use*
- *Emphasizing team spirit and staff participation*
- *Finding moves to improve, measure and supervise performance at institutional and individual levels*
- *Formation of corporate memory and adoption of acting with data"*

The adoption of the strategic planning approach in terms of public administration and its dissemination in the world have been tried to be implemented with the desire to gain these advantages. Each country's own internal dynamics have been effective in this process. Türkiye has encountered difficulties similar to other countries in the way of institutionalizing the strategic planning approach (Barati-Stec, 2015, pp. 5–14; Junjan, 2015; Hintea et al., 2015; Ravensway & Hamlin, 2015, pp. 55–70; Ruano, 2015; Schwab et al., 2017; Turc, Guenoun, Rodrigues, & Demirkaya, 2016).

The history of studies on the institutionalization of the planning approach in Türkiye is parallel to the reform studies in public administration. In particular, these reform efforts were implemented with legal regulations after the 2000s. Strategic plans gained a place in the Turkish public administration system with the regulations adopted and entered into force in this period. In this regard, the first step taken to ensure the continuity of strategic plans for public administration is the Public Financial Management and Control Law No. 5018. Article 9 of the law obliges public institutions to prepare strategic plans. In addition to this article, the obligation of local governments to prepare strategic plans is included in the Municipality Law No. 5393 and the Special Provincial Administration Law No. 5302. With Article 41 of the Municipality Law No. 5393, it has become mandatory for municipalities with a population of more than 50,000 and for Special Provincial Administrations with Article 31 of the Special Provincial Administration Law No. 5302, to prepare strategic plans (Official Gazette of the Republic of Türkiye, 2003; Official Gazette of the Republic of Türkiye, 2005).

Following these legal regulations in Türkiye, the State Planning Organization (DPT) prepared a guiding regulation that will contribute to the strategic plan preparation processes for public institutions. In line with the authority and duty granted by Law No. 5018, the "Regulation on Procedures and Principles Regarding Strategic Planning in Public Administrations" was prepared by DPT in 2006. Following this regulation, DPT assumed a guiding role in the strategic plan preparation processes of public institutions. As a result of the closure of the DPT in 2011 and the establishment of the Ministry of Development, the powers and duties of the DPT regarding strategic planning were transferred to the Ministry of Development (Official Gazette of the Republic of Türkiye, 2011).

As a result of the experiences gained in strategic planning, a new regulation was prepared by the Ministry of Development in 2018. With this regulation, the first regulation prepared in 2006 was abolished. The new regulation prepared by the Ministry of Development came into force under the name "Regulation on Procedures and Principles Regarding Strategic Planning in Public Administrations". However, the Ministry of Development, which prepared this regulation, was abolished as a result of the change in the government system in Türkiye and the restructuring of ministries. Following the transition to the presidential government system, the duties and powers of the Ministry of Development regarding strategic planning were transferred to the Strategy and Budget Directorate, established by Presidential Decree No. 13 (Official Gazette of the Republic of Türkiye, 2018).

Strategic plans are affected by legal changes and shaped according to the needs and visions of institutions. Medium- and long-term targets are determined by municipalities in order to rapidly increase the rate of urbanization and to provide solutions to problems such as environment, housing, health, and energy. Local administrators also attach importance to presenting their long-term strategic goals by analyzing their future opportunities and threats in order to use their limited resources more effectively and efficiently.

The fact that the concept of strategic planning has become important for cities is undoubtedly an issue that needs to be considered for Istanbul, which has the highest population in Türkiye, is at the center of the global financial system, and hosts international culture, art, and sports organizations. In this context, there is a need to determine what kind of strategy Istanbul has for environmental problems, and how it will combat the climate crisis, one of the important challenges faced by the world. It is necessary to see the most concrete form of this need of Istanbul in the strategic plans prepared by the IMM and analyze these plans.

Istanbul Metropolitan Municipality 2015–2019 Strategic Plan as a Tool for Challenge the Climate Crisis.

There is no direct statement of the climate crisis in the mission, vision, and values section, which constitute the components of the strategic plan prepared by the Justice and Development Party (AK Party) for the years 2015–2019. However, in the articles that reveal the mission and vision of the plan, the provision of urban services for the risks that threaten cities in the 21st century is mentioned. In this context, the mission of the plan is explained with the statement: "To provide services that will meet unrevealed needs in accordance with the local government approach of the 21st century, to keep the cultural identity of the city alive while perfecting municipal services with an institutional understanding" (Istanbul Metropolitan Municipality, 2014, p. 78).

The vision of the plan is expressed as "The local government of the brand city that makes life easier with sustainable and innovative solutions and produces global value in the name of urbanism and civilization." When looking at these sections of the plan, it is difficult to make a statement directly about combating the climate crisis. Despite this, the emphasis on the 21st century local government approach in the mission section of the plan indicates that an approach has been adopted to find solutions to the current problems faced by cities. The fact that the first concept highlighted in the vision section is sustainability indicates that environmental sensitivity forms the spirit of this plan (Istanbul Metropolitan Municipality, 2014, p. 78).

The values determined in the Strategic Plan of IMM for the years 2015–2019 are listed as "justice, trust, transparency, quality, efficiency, participation and leadership." These values are the headings that reveal the general principles to be adopted by the metropolitan municipality regarding the city's management process. There is no statement among the values that directly addresses the climate crisis (Istanbul Metropolitan Municipality, 2014, p. 79).

IMM has determined its strategic goals and targets by classifying urban services in line with the mission, vision, and values adopted and declared in its 2015–2019 Strategic Plan. Urban services are classified into eight main service areas and one general management area. Eight service areas have been determined as disaster, transportation, social support, urban and social order, zoning, culture, health and environmental management (Istanbul Metropolitan Municipality, 2014, p. 80).

When these service areas included in the plan are analyzed, it is seen that they are closely related to the Global Development Goals explained in the theoretical part of the study (Table 12.1). For example, the public policies to be developed with the Disaster Management service area are expected to contribute to the global goal of "Sustainable Cities and Living Spaces". Transportation Services, which is a secondary service area, also contributes to the global goal of "Sustainable Cities and Living Spaces". The tertiary service area of Social Support Services has the potential to correspond too many of the Global Development Goals. These are the global goals of "No Poverty, Zero Hunger, Quality Education, and Reduced Inequalities."

The fourth service area, Urban and Community Service, can be linked to the global goal of "Sustainable Cities and Habitats." The fifth service area, Zoning Management, is directly linked to the global goal of "Sustainable Cities and Habitats." The sixth service, Cultural Management, can be linked to the global goal of "Quality Education, Responsible Consumption and Production." The seventh service area, Health Services Management, addresses the global goals of "Healthy Individuals, Clean Water and Sanitation". Finally, the Environmental Management service is related to the global goals of "Clean Water and

Table 12.1. The Relationship Between the 8 Main Service Areas of IMM's Strategic Plan in 2015–2019 and Sustainable Development Goals.

| 8 Main Service Areas of Istanbul Metropolitan Municipality's Strategic Plan in 2015–2019 | Related Sustainable Development Goals |
|---|---|
| **Disaster management** | Sustainable cities and living spaces |
| **Transportation services** | Sustainable cities and living spaces |
| **Social support services** | No poverty, zero hunger, quality education, and reduced inequalities |
| **Urban and community services** | Sustainable cities and habitats |
| **Zoning management** | Sustainable cities and habitats |
| **Cultural management** | Quality education, responsible consumption, and production |
| **Health services management** | Good health and well-being, clean water, and sanitation |
| **Environmental management** | Clean water and sanitation, affordable and clean energy, climate action, life below water, life on land |

Sanitation, Affordable and Clean Energy, Climate Action, Life Below Water, Life on Land" and can be directly addressed with the topic of climate change, which is the subject of this research.

When the strategic goals and targets that will contribute to the fight against the global climate crisis in the 2015–2019 Strategic Plan of IMM are examined, it is seen that the subgoals determined in different service areas have the potential to contribute to the fight against the climate crisis at different rates. In this context, "Increasing the Number of Low Emission Public Transportation Vehicles" target has been set in the fifth of the sub-targets related to the target for Transportation Services Management. Considering that global climate change is mainly caused by air pollution and carbon emissions, this target stands out directly.

The tangible counterpart of IMM's global climate change targets for this period can be seen in the targets under target number eight, Environmental Management Services. The first target is "Improving Waste Management Practices," the second is "Disseminating Environmental Protection Practices," the third is "Improving Cooperation with Stakeholders in Environmental Management," the fourth is "Improving Practices for the Prevention of Marine Pollution," the fifth is "Carrying out Studies on the Maintenance and Protection of Monumental Trees that have Witnessed History," the sixth target is "Enabling and Disseminating Activities for Renewable Energy and Energy Efficiency," and the seventh target is "Disseminating Green Areas." Except for the fifth target, all other targets are of strategic importance in the fight against the global climate crisis.

In the strategic plan prepared by IMM, the total amount of resources envisaged for strategic purposes that will contribute to the fight against the climate crisis has been determined as nine billion 110 million 772 thousand 260 Turkish liras. The amount of resources allocated to all urban services for five years in the entire strategic plan is 64 billion 283 million 162 thousand 191 Turkish liras. Therefore, IMM has allocated 14% of its resources to the services it will implement to combat climate change, directly or indirectly.

It is seen that the six targets determined by the IMM in the fight against the climate crisis in this period, five of which are environment and one of which is transportation services, are included in the strategic plan with different indicators and are being followed. As an indicator for the target of "Increasing the Number of Low-Emission Public Transportation Vehicles" in transportation services, it is aimed to increase the number of low-emission minibuses, taxis, and minibuses from 1,500 to 6,500 in five years. It is envisaged that IMM will not need any financial resources to achieve this goal. It is planned that the renewal in question will be carried out by the vehicle owners themselves.

The target of "Improving Waste Management Practices" determined in environmental services was followed with 11 indicators. These indicators are the daily amount of solid waste disposed of in regular landfills, the amount of compost produced, the number of new landfills, the amount of medical waste sterilized, the amount of electricity produced from methane gas, the number of electronic units entering and processing the facility, and training on waste management and city cleaning. It was determined as the number of people assigned, the amount of waste processed in the integrated mechanical biological process and recovery facility, the amount of waste transported from transfer stations to the landfill, the number of new solid waste transfer stations, and the number of completed projects.

The target of "Making Environmental Protection Practices Common" was followed with 14 indicators. These indicators include: the main artery area cleaned throughout Istanbul, the completion rate of noise maps, the number of personnel trained on environmental pollution and inspection, the number of people reached within the scope of developing environmental awareness, the number of coal sampling and inspections, the number of vehicles equipped with GPS for excavation inspection, daily inspection hours for marine pollution, number of measured pollutants, number of accredited parameters, completion rate of the Istanbul Climate Change Action Plan, timely resolution rate of complaints from the White Table (Beyaz Masa), satisfaction rate of applicants with the response time of applications from the White Table, solution offered to applications from the White Table, satisfaction rate of applicants, and capacity of the area allocated for excavation storage.

The target of "Strengthening Cooperation with Stakeholders in Environmental Management" was monitored with one indicator. This indicator was determined as the number of landscape intermediate staff trained. The target of "Improving Practices for the Prevention of Marine Pollution" was monitored with nine indicators. These indicators are as follows: the amount of petroleum-derived waste collected from marine vessels, the amount of sewage waste collected from marine vessels, the amount of garbage waste collected from marine vessels, the amount of petroleum-derived waste recovered from marine vessels, the amount of sediments from creek mouth and bay dredging activities, the amount of solid waste collected from the sea surface, the amount of waste collected from coasts and beaches, the number of trainings organized to increase awareness of marine cleanliness, and the length of the stream rehabilitated.

The target of "Enabling and Disseminating Activities for Renewable Energy, Energy Efficiency" was monitored with six indicators. These indicators were determined as the number of places where the lighting system was renewed during the year, periodic maintenance and repair of generators within IMM, the number of people reached within the scope of awareness raising activities on energy efficiency, monetary savings achieved in energy purchase in the free market, the number of IMM facilities where solar energy system will be implemented during the year, the number of periodic maintenance of elevators, escalators and disabled elevators in Metrobus, and upper and lower passages under the responsibility of IMM.

Finally, the goal of "Making Green Areas Expand" was followed with five indicators.[1] These indicators were determined as the number of newly established children's play groups, the number of newly established fitness groups, the amount of new green areas, the amount of maintained green areas, and the number of flowers planted to increase visual richness.

## Istanbul Metropolitan Municipality 2020–2024 Strategic Plan as a Tool for Challenge the Climate Crisis

The mission, vision, and values of the IMM Strategic Plan, created by the Republican Party People (CHP) for the 2020–2024 period, contain an important emphasis on combating the global climate crisis. In this context, the emphasis on "green city" stands out as the vision of this period. However, in the mission section, there are statements such as "To provide an accessible service for an Istanbul that lives 24 hours a day with its productive, cultural and social life by protecting all the values of the city with a new generation municipalism approach" (Istanbul Metropolitan Municipality, 2019, p. 10). Even if these statements show that there is no emphasis on the environment or climate change in the mission of the plan, the vision of the plan focuses on creating an imagined Istanbulite profile by solving Istanbul's green space problem by using the expressions "a fair, green and creative city, a happy Istanbulite" (Istanbul Metropolitan Municipality, 2019, p. 10).

The values section of the IMM's 2020–2024 plan includes the titles "justice, transparency and accountability, ethics, inclusiveness, innovation, environmental sensitivity, merit, participation, human orientation and agility". Among these

[1]12 indicators have been determined for the goal of Expanding Green Areas. However, since it was understood that these indicators were duplicates in the plan, only unique indicators were included in the text.

headings, the value of "sensitivity to the environment" adopted for combating the climate crisis is explained with the statements "Carrying out municipal services without depriving future generations of today's environmental assets and increasing the green areas of Istanbul accessible to everyone". The most important statement in terms of combating the climate crisis is the responsibility toward future generations (Istanbul Metropolitan Municipality, 2019, p. 11).

In the IMM 2020–2024 plan, urban services are defined by eight themes. These themes are categorized as "Accessible Istanbul, Environmentally Sensitive, Producing Istanbul, Sharing Istanbul, Living Istanbul, Unique Heritage, Financial Sustainability, Participatory and Innovative Management" (Istanbul Metropolitan Municipality, 2019, pp. 12–14). These themes, which determine the services to be provided to Istanbul, are directly or indirectly related to the SDGs set by the UN (Table 12.2).

Within the first theme, "Accessible Istanbul", services such as urban planning, transportation, modern infrastructure, urban informatics, emergency health, disaster and emergency response, and urban transformation have been identified.

Table 12.2. The Relationship Between the 8 Main Themes of IMM's Strategic Plan in 2020–2024 and Sustainable Development Goals.

| 8 Main Themes of Istanbul Metropolitan Municipality's Strategic Plan in 2020–2024 | Related Sustainable Development Goals |
|---|---|
| **Accessible Istanbul** | Sustainable cities and living spaces |
| **Environmentally sensitive** | Clean water and sanitation, affordable and clean energy, sustainable cities and habitats, climate action, life below water, life on land |
| **Producing Istanbul** | Decent work and economic growth, industry, innovation, and infrastructure |
| **Sharing Istanbul** | No poverty, zero hunger, reduced inequalities |
| **Living Istanbul** | Good health and well-being, clean water and sanitation, responsible consumption and production |
| **Unique heritage** | None |
| **Financial sustainability** | Peace, justice, and strong institutions |
| **Participatory and innovative management** | Peace, justice, and strong institutions, partnerships for the goals |

These services are directly linked to the 11th target of the SDGs, namely "Sustainable Cities and Living Spaces." Within the second theme, "Environmentally Responsible", services such as green space, sea, air, water management, energy and energy efficiency, waste management have been identified. These services are also related to many of the SDGs. These are the sixth goal "Clean Water and Sanitation," the seventh goal "Affordable and Clean Energy," the 11th goal "Sustainable Cities and Habitats," the 13th goal "Climate Action," the 14th goal "Life Below Water," and the 15th goal "Life on Land".

Within the third theme, "Producing Istanbul", services such as employment, tourism, on-site agriculture, entrepreneurship, investment office, and vocational trainings have been identified. These services are in line with the eighth SDG "Decent Work and Economic Growth" and the ninth goal "Industry, Innovation and Infrastructure." "Social Inclusive Services, Social Support Services, Social Assistance" services are mentioned under the fourth theme "Sharing Istanbul." It is seen that these service areas are linked to the first goal of "No Poverty," the second goal of "Zero Hunger," and the 10th goal of "Reduced Inequalities."

The fifth theme, "Living Istanbul," includes health, fight against vectors, culture and arts, sports, events, international organizations, sports facilities, animal rights and life, veterinary services, inspections, permits and licenses, and consumer rights services. These services are related to the third SDG, "Good Health and Well-Being," the fourth goal, "Clean Water and Sanitation," and the 12th goal, "Responsible Consumption and Production." The sixth theme, "Unique Heritage," includes the services of our tangible/intangible historical assets. It is difficult to associate the titles in this theme with any of the SDGs.

The seventh theme, "Financial Sustainability," includes IMM's budget, savings, new business models, alternative financing sources, and effective asset management services. These services are related to the 16th SDG, "Peace, Justice, and Strong Institutions." The eighth theme "Participatory and Innovative Management" includes service areas such as good governance, occupational health, organizational development, subsidiary management, operational excellence, and stakeholder orientation. These service areas are related to the 16th goal "Peace, Justice, and Strong Institutions" and the 17th goal "Partnerships for the Goals".

It is seen that some of the strategic goals and targets prepared in accordance with the urban services determined in the 2020–2024 Strategic Plan of IMM are aimed at combating the global climate crisis. In this context, it is seen that the four targets determined under the goal number 3 "Strengthening Sustainable Environment and Energy Management" are the headings that will contribute to climate change. These goals are as follows: Increasing the Recycling Rate by Improving Waste Management Practices; Enabling and Disseminating Activities for Renewable Energy, Energy Efficiency, and Lighting Facilities; Creating Sustainable Green Areas and Increasing the Amount of Active Green Areas Per Capita; and Protecting the Environment by Expanding the Fight against Climate Change.

In the strategic plan prepared by IMM, the total amount of resources envisaged for strategic purposes that will contribute to the fight against the climate crisis has been determined as 15 billion 123 million 232 thousand 501 Turkish liras. The amount of resources allocated to all urban services for five years in the entire strategic plan is 143 billion 330 million Turkish liras. Therefore, IMM has allocated 10.5% of its resources to the services it will implement to combat climate change, directly or indirectly.

It is seen that the four targets determined by the IMM in the fight against the climate crisis in this period, five of which are environment and one of which is transportation services, are included in the strategic plan with different indicators and are being followed. The target of "Increasing the Recycling Rate by Improving Waste Management Practices" was followed with three indicators. These indicators were determined as the amount of electrical energy produced from waste, the rate of waste processed in recycling facilities, and the corporate waste recycling rate.

The target of "Enabling and Disseminating Activities for Renewable Energy, Energy Efficiency, and Lighting Facilities" was followed with six indicators. These include: the rate at which lighting facilities are kept lit, the rate at which municipal service buildings are converted into green buildings, the number of people receiving annual energy efficiency awareness training, the rate at which the municipality's electricity consumption is covered by renewable energy, the rate of lighting facilities that can be remotely controlled with automation, and the rate at which renewable energy is installed.

The target of "Creating Sustainable Green Areas and Increasing the Amount of Active Green Areas per Capita" was followed with five indicators. These indicators include: the amount of green areas maintained annually, the number of trees planted annually, the amount of active green areas per person per year, the number of trees with annual civil-cultural maintenance and restoration, the number of newly established fitness and children's play groups annually, the total number of newly built amount of green space.

The target of "Protecting the Environment by Expanding the Fight against Climate Change" was followed with three indicators. These indicators are the annual number of events carried out to increase environmental awareness, the total number of inspections carried out within the scope of environmental management, and the update rate of the urban greenhouse gas emission inventory.

## Conclusion

The global climate crisis, which knows no boundaries of region or country, causes significant problems in cities as a result of population movements and economic activities. Increasing global average temperatures and the resulting rise in sea level, extreme weather events such as floods, droughts and storms, and a corresponding increase in the number and effectiveness of disaster events, and the prevalence of public health problems are some of the problems in cities. These problems undoubtedly affect many systems within urban life. Urban infrastructure, social services, people's health and well-being, and the structure of the housing stock are examples of these. This situation in cities does not change the fact that climate change has an impact on cities, as well as the fact that economic and human activities in cities are related to climate change. From this perspective, climate change is a two-dimensional topic for cities. New conditions arising as a result of climate change affect cities negatively, and activities carried out in cities may be among the factors that cause climate change. As a result, cities are affected by climate change in some cases, and in other cases, cities are the source of climate change.

Climate change is defined by the UN as long-term changes in the climate caused directly or indirectly by human influence. As can be understood from the definition, climate change is one of the important problems that disrupt the global atmospheric composition that the world faces today as a result of human activities. The biggest example of this situation is the share of cities in the world's carbon dioxide emissions. Seventy-five percent of carbon dioxide emissions occur as a result of activities carried out in cities. Especially in cities, the density of the industrial sector, means of transportation, and settlements affects this situation.

This situation, where cities are intertwined with global climate change, is undoubtedly also valid for metropolitan areas. Metropolitan city governments take global climate change into account when determining their medium- and long-term strategic goals and objectives. Accordingly, cities determine their strategic goals and objectives. This situation remains valid for Istanbul, Türkey's foremost city in terms of population, economic activity, and global connections. IMM, one of the main institutions responsible for finding solutions to the problems of Istanbul as a metropolis, has also included climate change on its agenda in the five-year strategic plans it has prepared in different periods.

The strategic plans prepared by IMM for 2015–2019 under the AK Party government and for 2020–2024 under the CHP government attract attention as important issues, as well as the compliance with the SDGs adopted by the UN and the contribution of against to adaptation global climate change. When the strategic plan prepared by AK Party in 2014 is examined from different dimensions, it is seen that although the plan was prepared before the decision taken by the UN for the SDGs, its compliance with these goals reveals the success of the plan prepared in the relevant period. The plan prepared by CHP for the period 2015–2019 includes the SDGs "Sustainable Cities and Communities, No Poverty, Zero Hunger, Quality Education, Reduced Inequalities, Good Health and Well-Being, Clean Water and Sanitation, Affordable and Clean Energy, Climate Action, Life Below Water, Life on Land." It is very important that it is compatible with these goals. Apart from this, the fact that all six of its strategic goals and objectives are followed up with public services aimed at combating global climate change is an example of the importance that IMM attaches to this issue. In addition, the resource allocated by IMM to accept these services as indicators corresponds to 14% of the resource foreseen for all services in the five-year plan, indicating that a significant resource transfer will be made to areas that may be related to climate change for the years 2015–2019.

In the strategic plan prepared by CHP for the period 2020–2024, it is seen that the decision taken by the UN regarding the SDGs in 2015 was effective. As a matter of fact, care was taken to ensure that the strategic goals and targets determined throughout the plan were compatible with the SDGs, and even a table showing this harmony was prepared. In this context, even though the 2020–2024 Strategic Plan of IMM under CHP is a plan compatible with the SDGs, it lags behind the plan prepared by AK Party for the 2015–2019 period in determining concrete indicators for combating global climate change and enriching. As a matter of fact, this situation can be understood from the determined indicators.

## References

Ayva, C., Dutucu, A. A., & Ustaoğlu, B. (2023). İklim Değişikliğinin Su Kaynaklarına Etkisi ve Uyum Önerileri: Kirazdere Havzası Örneği. *Fırat Üniversitesi Sosyal Bilimler Dergisi*, *33*(1), 47–64.

Barati-Stec, L. (2015). Strategic planning in Hungarian municipalities. *Transylvanian Review of Administrative Sciences* (Special Issue), 5–14.

Berry, F. S. (1994). Innovation in public management: The adoption of strategic planning. *Public Administration Review*, *54*(4), 322–330.

Çelik, S., Bacanlı, H., & Görgeç, H. (2008). *Küresel İklim Değişikliği ve İnsan Sağlığına Etkileri*. Telekomünikasyon Şube Müdürlüğü.

Çelik, A., & Küçük, A. (2020). Tüketim Toplumunun Çevre Sorunlarına Etkileri. *Ecoharran Harran Üniversitesi İİBF Dergisi*, *4*(5), 1–22.

Demir, A. (2009). Küresel İklim Değişikliğinin Biyolojik Çeşitlilik ve Ekosistem Kaynakları Üzerine Etkisi. *Ankara Üniversitesi Çevrebilimleri Dergisi*, *1*(2), 37–54.

Demirkaya, Y. (2015). Strategic planning in the Turkish Public Sector. *Transylvanian Review of Administrative Sciences* (Special Issue), 15–29.

Demirkaya, Y. (2019). Kamu Yönetiminde Planlamayı Keşfetmek: Stratejik Planlama. *V. Stratejik Kamu Yönetimi Sempozyumu Tam Metin Bildiri Kitabı* (pp. 9–14).

Favoreu, C., Carassus, D., & Maurel, C. (2015). Strategic management in the public sector: A rational, political or collaborative approach? *International Review of Administrative Sciences*, *82*(3), 1–19.

Gernaat, D. E., Boer, H. S., Daioglou, V., Yalew, S. G., Müller, C., & Vuuren, D. P. (2021). Climate change impacts on renewable energy supply. *Nature Climate Change*, (11), 119–125.

Gunawan, J., Permatasari, P., & Tilt, C. (2019). Sustainable development goal disclosures: Do they support responsible consumption and production? *Journal of Cleaner Production*, (246), 1–12.

Hintea, C. E., Profiroiu, C. M., & Ticlau, T. C. (2015). Strategic planning and public management reform: The case of Romania. *Transylvanian Review of Administrative Sciences* (Special Issue), 30–44.

Istanbul Metropolitan Municipality. (2014). Strategic plan 2015-2019. http://www.sp.gov.tr/upload/xSPStratejikPlan/files/FWj9k+Istanbul_BB_stratejikplan15_19.pdf. Retrieved from February 13, 2024.

Istanbul Metropolitan Municipality. (2019). Strategic plan 2020-2024. http://www.sp.gov.tr/upload/xSPStratejikPlan/files/4mFEW+iBB-STRATEJIK-PLAN-2020-2024.pdf. Retrieved from February 13, 2024.

Junjan, V. (2015). Strategic planning in local governments in Europe: 'Where do we go now? *Transylvanian Review of Administrative Sciences* (Special Issue), 45–54.

Kavas, A. (2016, January 5). Sürdürülebilir kalkınma, geniş tabanlı, evrensel bir anlayışla, küresel dünyanın gündemine yerleşiyor. https://www.tepav.org.tr/: https://www.tepav.org.tr/tr/blog/s/5470/Surdurulebilir+kalkinma_+genis+tabanli_+evrensel+bir+anlayisla_+kuresel+dunyanin+gundemine+yerlesiyor. Retrieved from February 13, 2024.

Kayhan, F. E., Kaymak, G., Tartar, Ş., Akbulut, C., Esmer, H. E., & Ertuğ, N. D. (2019). Küresel Isınmanın Balıklar ve Deniz Ekosistemleri Üzerine Etkileri. *Erciyes Üniversitesi Fen Bilimleri Dergisi, 31*(3), 128–134.

Keskin, T. (2019). *Enerji Politikalarının İklim Değişikliği ile Mücadeledeki Yeri.* WEglobal.

Menga, M. (2023, July 14). Land and climate change: A close connection. https://www.climateforesight.eu/articles/land-and-climate-change-a-close-connection/. Retrieved from July 24, 2024.

Official Gazette of the Republic of Türkiye. (2003). *Public Financial Management and Control Law No. 5018.*

Official Gazette of the Republic of Türkiye. (2005). *Special Provincial Administration Law No. 5302.*

Official Gazette of the Republic of Türkiye. (2011). *On the Organization and Duties of the Ministry of Development Decree Law No. 641.*

Official Gazette of the Republic of Türkiye. (2018). *Presidential Decree No. 13.*

Peşkiricioğlu, N. (2016). 2030 Sürdürülebilir Kalkınma Hedefleri: Küresel Verimlilik Hareketine Doğru. *Anahtar Dergisi, 28*(355), 4–9.

Pollitt, C., & Bouckaert, G. (2011). *Public management reform: A comparative analysis - new public management, governance, and the Neo-Weberian State.* Oxford University Press.

Ravensway, J. V., & Hamlin, R. (2015). *Strategic planning in U.S. municipalities. Transylvanian Review of Administrative Sciences* (Special Issue), (pp. 55–70).

Ruano, J. M. (2015). Local strategic planning in Spain. A case study. *Transylvanian Review of Administrative Sciences* (Special Issue), 71–85.

Sand, P. H. (2007). The evolution of international environmental law. In D. Bodansky, J. Brunnée, & E. Hey (Eds.), *The Oxford Handbook of international Enviromental Law* (pp. 31–39). Oxford University Press.

Schwab, C., Bouckaert, G., & Kuhlmann, S. (2017). *The future of local government in Europe: Lessons from research and practice in 31 countries.* Nomos.

Sukkar, E. (2021). Climate change and its impact on lung health: A focus on Europe. https://impact.economist.com/perspectives/health/climate-change-and-its-impact-lung-health-focus-europe?utm_medium=cpc.adword.pd&utm_source=google&ppccampaignID=18151738051&ppcadID=&utm_campaign=a.22brand_pmax&utm_content=conversion.direct-response.anonymous&gad_source=1&gclid=Cj0KCQiAxOauBhCaARIsAEbUSQSCzfS0xsyhA-TcdD2BTJnI0diQ_FNNQfO9MiTd815GlCoWWM58HOEaAnt9EALw_wcB&gclsrc=aw.ds. Retrieved from July 24, 2024.

Turc, E., Guenoun, M., Rodrigues, M., & Demirkaya, Y., (2016). Performance Management. In S. Kuhlmann, & J. Bouckaert içinde (Eds.). *Local public sector reforms in times of crisis: National Trajectories and International Comparisons.* Palgrave.

United Nations. (2024c). *Conferences: Environment and sustainable development.* https://www.un.org/en/conferences/environment/rio1992. Retrieved from March 12, 2024.

United Nations. (1987). Of the world Commission on Environment and development: Our common future. https://sustainabledevelopment.un.org/content/documents/5987our-common-future.pdf. Retrieved from March 12, 2024.

United Nations. (2015a). UN.org. https://www.un.org/millenniumgoals/: https://www.un.org/millenniumgoals/bkgd.shtml. Retrieved from February 13, 2024.

United Nations. (2015b). *Department of economic and social affairs sustainable development*. sdgs.un.org. https://sdgs.un.org/2030agenda. Retrieved from February 13, 2024.

United Nations. (2015c). *The global goals*. https://www.globalgoals.org/: https://www.globalgoals.org/. Retrieved from February 13, 2024.

United Nations. (2015d). Population 2030. https://www.un.org/en/development/desa/population/publications/pdf/trends/Population2030.pdf. Retrieved from February 13, 2024.

United Nations. (2024a). History of the United Nations. https://www.un.org/en/about-us/history-of-the-un. Retrieved from March 12, 2024.

United Nations. (2024b). *Conferences: Habitat*. https://www.un.org/en/conferences/habitat. Retrieved from March 12, 2024.

United Nations. (2024d). What is the Kyoto protocol? https://unfccc.int/kyoto_protocol#:~:text=In%20short%2C%20the%20Kyoto%20Protocol,accordance%20with%20agreed%20individual%20targets. Retrieved from March 12, 2024.

Wheeler, T., & Braun, J. V. (2013). Climate change impacts on global food security. *Science*, (341), 508–513.

Chapter 13

# The European Union's Policies and Role in Tackling Climate Change in the Context of the European Green Deal

*Kübra Ecer*[a,1] *and Oğuz Güner*[b,2]

[a]The Republic of Türkiye Directorate of Communications, Turkey
[b]Ankara Yildirim Beyazit University, Turkey

## Abstract

In recent years, many nations have accelerated efforts to increase public awareness on environment and reduce the negative impact of the climate change. The European Union (EU) has been one of the most active international actors in environmental protection, implementing policies focused on providing a sustainable future for its citizens and protecting natural resources. The European Green Deal (EGD) which is a comprehensive plan with the goal of making the EU a climate-neutral zone by 2050 includes strategies, policies, and measures in numerous sectors such as economy, energy, transportation, agriculture, and industry. The EGD's targets include reducing greenhouse gas emissions by at least 55% by 2030 compared to 1990 levels, increasing the share of renewable energy at least 42.5% within the EU by 2030, promoting circular economy, and encouraging the use of low-emission vehicles (European Commission, 2020a). The Deal is set to transform the European economy into a more sustainable and low-carbon economy and fight against climate change all around the Europe and partner countries. The EU is committed to provide financial support for the actions required to address the root causes of climate change and environmental issues. This chapter examines the initiatives adopted by the EU to mitigate environmental issues and the effects of climate change in the context of the EGD. The study's objective is to assess the EU's efforts to help bring about a

[1]https://orcid.org/0000-0002-6519-2803
[2]https://orcid.org/0000-0003-1619-579X

**The Social Consequences of Climate Change, 163–185**

**doi:10.1108/978-1-83797-677-520241013**

more sustainable environment in which negative impacts of the climate change and the release of harmful gases are significantly reduced.

*Keywords*: European Union; climate change; European Green Deal; circular economy; environmental economy

## Introduction

The deteriorating environmental situation has been prompting many international actors, including the EU, to take action to mitigate climate change. Turning the tide on climate change has become a priority for the EU to ensure protecting natural resources and providing a sustainable future for its citizens.

The European Green Deal (EGD), which is a comprehensive plan aiming at making the EU a climate-neutral zone by 2050, presents a roadmap for numerous sectors such as economy, energy, transportation, agriculture, and industry. The EGD targets reducing greenhouse gas emissions by at least 55% by 2030 compared to 1990 levels, increasing the share of renewable energy at least 42.5% within the EU by 2030, promoting circular economy, and encouraging the use of low-emission vehicles. The Deal necessitates transformation of the European economy into a more sustainable economy. Within the framework of the Deal, the EU is committed to provide financial support for steps to be taken to address the climate issue. The European Commission has committed to mobilizing at least €1 trillion in sustainable investments over the next 10 years in order to meet the objectives set forth in the EGD (European Commission, 2020b).

This study examines the initiatives adopted by the EU to mitigate environmental issues and the effects of climate change in the context of the EGD. The study is intended to assess the EU's efforts of climate action and how it makes efforts to help build a more sustainable environment where greenhouse gas emissions are significantly reduced. To this end, the first chapter offers an overview of EGD and steps taken with regard to the plan. The second chapter evaluates the EU's objectives and commitments for climate action with focus on the reduction of greenhouse gas emissions and circular economy. The final chapter dwells on the global relevance of the EU's climate policy.

## Overview of the EGD

Climate change and its devastating impacts on earth including floods, deadly heat, and fires have long been ringing alarm bells. To stave off catastrophic warming, international organizations call for immediate action to reverse the trend. The EU announced the EGD on December 11, 2019, which has an overarching aim of transforming Europe into the first climate-neutral continent by 2050 (European Commission, 2019d).

It is a comprehensive plan charting strategies, policies, and measures in numerous sectors such as economy, energy, transportation, agriculture, and industry. The EGD aims at reducing greenhouse gas emissions by at least 55% by 2030, increasing the share of renewable energy at least 42.5%, promoting circular economy, and encouraging the use of low-emission vehicles (European Commission, 2020a).

The targets outlined in the EU's deal are legally binding and cover all key sectors of the economy such as energy, agriculture, transport, and industries of steel and cement. Generating investments in the green transition, putting a price on pollution, and social support for citizens are essential objectives in the package (European Commission, 2020a). In relation to the EGD, President of the European Commission Ursula von der Leyen highlighted that the EGD presents the EU's new growth strategy and paves the way for transformation of all aspects of life. (European Commission, 2022, para. 2). The EGD sets outs priorities of the EU to achieve its objective of becoming climate neutral by 2050. The EGD encompasses various issues including fight against climate change, strategies for climate neutrality, and transformation of economy. The EGD addresses steps to be taken to reach the objectives regarding climate neutrality. Therefore, an in-depth overview of the key concepts is necessary to provide insight into the EU's policies.

### *Path Toward Climate Neutrality*

The transition to climate neutrality has prompted the EU to introduce various plans and strategies to enable the sustainable transformation of the economy. Resource extraction and processing comprises half of total greenhouse gas emissions. It substantially causes biodiversity loss and water stress (EUR-Lex, 2020). For a climate-neutral and resource-efficient economy, the European Commission adopted the new circular economy action plan (CEAP) in March 2020 which is an essential part of the EGD (European Commission, 2020b). It was highlighted that the EU's transition to a circular economy would help reducing the waste of natural resources and enable sustainable growth. CEAP focuses on production processes, sustainable consumption, and preventing of waste of resources.

To navigate the path toward EU's 2030 climate targets, the Fit for 55 legislation was prepared as an essential part of the EGD. Fit for 55 refers to the EU's target of reducing net greenhouse gas emissions by at least 55% by 2030 (European Council, 2021c). With the completion of the Fit for 55 packages in October 2023, the EU had legally binding targets which cover all key sectors in the economy (European Commission, 2021c).

With the package, proposals including EU Emissions Trading System (ETS) reform, New EU ETS for building and road transport fuels, Social Climate Fund, Effort Sharing Regulation, Regulation on Land Use, Forestry and Agriculture (LULUCF) were adopted (European Commission, 2023c).

A comprehensive transformation of all key sectors in a sustainable manner is required to reach the objectives of the EU's climate action plan. Emissions reduction targets in key sectors, social support for citizens and small businesses, and updating of ETS to cap emissions stand out as priorities for the EU countries (European Commission, 2021b).

While the EU is taking decisive steps toward its goal of climate neutrality, its partners and allies may need to keep abreast of these comprehensive changes. The

EU takes measures to ensure a level playing field for European countries so that they will not be negatively affected by the policies of non-EU countries which do not adopt decisive climate action. Fearing the risk of "carbon leakage," the EU introduced new plans as part of its Green Deal Plan. It is explained that carbon leakage occurs when European companies opt to produce carbon-intensive goods outside of the EU where climate policies might be less drastic (European Commission, 2019c). To prevent carbon leakage, the EU introduced Carbon Border Adjustment Mechanism (CBAM) and required payment for the carbon emitted during the production of carbon-intensive goods (European Commission, 2023c). The Mechanism entered into application in its transitional phase on October 1, 2023, and the first reporting period for importers ends on January 31, 2024 (European Commission, 2023c, 2024d). It is highlighted that to prevent carbon leakage, CBAM will first be implemented for the imports of some specific goods such as cement, aluminum, fertilizers, iron and steel (European Commission, 2023c).

Accordingly, the EU importers will be expected to purchase CBAM certificates, equal to the carbon price that would have been paid if the goods had been manufactured in compliance with EU carbon pricing regulations (European Commission, 2024e, p. 8). Furthermore, a non-EU producer's cost will be completely subtracted from the CBAM obligation if they have previously paid the carbon price in a third country for the embedded emissions of the imported goods (European Commission, 2024e). The EU highlights that this Mechanism will encourage non-EU countries to transform the process of their productions.

### *Climate Action in the Context of Green Deal*

As the impact of the climate change has been unprecedentedly felt around different parts of the world, it is brought on top of the world's agenda. The indicators highlight the need for immediate reductions in emissions to provide a sustainable future and livable planet for the future generations. There have been efforts to pursue common action and manage impacts of the climate change.

The Paris Agreement is a significant step stressing the urgency of climate action and provides a roadmap for tackling it. A legally binding global climate agreement, the Paris Agreement was adopted in 2015 during the UN Climate Change Conference (COP21) in Paris. Its main goals are to keep global average temperature increases to well below 2°C above preindustrial levels and to work toward keeping increases to 1.5°C above preindustrial levels (United Nations, 2015b para. 2). Accordingly, the parties are expected to present their action plans, which are named nationally determined contributions (NDCs), and enhance their efforts to reduce emissions.

In relation to the climate change's impact, the UN Secretary-General António Guterres highlights that "The era of global warming has ended; the era of global boiling has arrived" (Niranjan, 2023, para. 1). As the world leaders convened in New York in September 2023 for the 78th session of the UN General Assembly, Secretary-General Guterres called for action to revive the Sustainable Development Goals (SDGs). The SDGs include a wide range of objectives from ending poverty to protecting the earth's environment and climate. The "Climate Action" is one of the UN's SDGs. The UN stressed that if required measures are not taken, the number of people who lack sufficient water is expected to rise from 3.6 billion today to more than 5 billion by 2050 (Watts, 2018). According to the report published by the World Bank in 2021, more than 216 million people might migrate because of climate change's negative impact on their lives by 2050 (The World Bank, 2023). Most of this population change is predicted to take place in sub-Saharan Africa, South Asia, and Latin America.

Aiming to take concrete steps regarding the climate action, the European Climate Law was adopted by the EU Council in 2021, setting up a framework to achieve climate neutrality in 2050. The law encompasses tracking progress made by member states, reviewing progress made every 5 years, and enabling the Commission to issue recommendations to member states, which do not take necessary steps for climate neutrality target (European Commission, 2021b).

In 2023, the Commission evaluated the progress regarding the framework of the climate neutrality target in 2050, and the assessment was published at the 2023 Climate Action Report. Accordingly, the Commission highlighted that the "EU's climate neutrality objective appears to be insufficient," even though greenhouse gas emissions continued to fall (European Commission, 2024e, para. 15). It was also underlined that the EU has made progress on adaptation to climate change, yet the progress was not at the same level across areas. Member states were encouraged to take steps regarding the areas of governance, monitoring, and reporting to mitigate social and economic vulnerabilities to the climate-related risks.

As climate change requires global response, the EU engages with international actors to support efforts to mitigate challenges posed by climate change. The EU's members have all signed and ratified the Paris Agreement; the Union seeks to implement actions and take necessary steps for transition toward sustainable climate-neutral economies (European Council, 2019e). Furthermore, the European Commission presented an intermediate 2040 climate target in February 2024 which envisions reducing net greenhouse gas emissions by 90% by 2040 compared to 1990 levels (European Commission, 2024e). It was stressed that reducing net emissions by 90% by 2040 will put the EU on track toward climate neutrality (European Commission, 2024e). As European countries enhance their efforts for transition to climate-neutral economies, the EU promises to protect the most vulnerable citizens and small businesses with the green transition, aiming to leave no one behind (European Commission, 2024e). The Social Climate Fund is set to dedicate €65 billion from the EU budget and over €86 billion in total for supporting people and businesses.

To better address the climate change, all sectors have been put onto spotlight, and strategies are drawn to reduce the emissions. By 2035, the new $CO_2$ regulations require all newly registered automobiles and vans in Europe to be zero emissions (European Commission, 2024e). The intermediate stage toward zero emissions will be a 55% reduction in automobile emissions by 2030. The EU is also taking steps to create the infrastructure that is needed to charge zero-emission vehicles. In the aviation sector, carbon pricing is applied in European Economic Area (EEA). From 2024, other regions will be included as well.

The EU introduced Biodiversity Strategy for 2030 aiming at protecting nature and reversing the degradation of ecosystems. Investing in restoring nature and biodiversity is considered a cheap and effective solution to absorbing carbon (European Commission, 2020a). The ambitious and long-term Strategy is an essential part of the Green Deal. The Biodiversity Strategy focuses on building resilience of communities against unprecedented challenges including the impacts of climate change, forest fires, and food insecurity (European Commission, 2020a). It is emphasized that one million species are in danger of going extinct and that the number of wild species worldwide has decreased by 60% in the past 40 years (European Commission, 2020a). Restoring forests and constructing new green areas can help ease the impacts of climate change, which is exacerbated by biodiversity loss. At least, 30 percent of Europe's land and 30 percent of its sea will be protected areas according to the Biodiversity Strategy. The Strategy will also unlock funding for biodiversity that will ensure the necessary transformative change. By 2030, the use of pesticides will be reduced by 50%, and 3 billion trees will be planted.

### *Energy Transition and the Green Deal*

Increased energy efficiency and higher shares of renewable energy are required to reduce gas emissions by at least 55% by the year 2030. The developments in international relations particularly the war in Ukraine have disrupted energy markets and energy security. The EU, which accelerated its efforts to wean itself from the Russian gas, unveiled REPowerEU plan in May 2022. REPowerEU puts a spotlight on deploying more renewable energy, saving energy, and diversifying energy supplies (European Commission, 2022). The EU's renewable energy target for 2030 is 42.5%, and the Union is set to improve energy efficiency by 11.7% by 2030 (European Commission, 2018). Russia–Ukraine war has highlighted that the EU needs to diversify its energy resources to ensure its energy security. Through the REPowerEU, the EU aims to diversify its energy sources by entering into pipeline import agreements with other third parties, investing in shared liquefied natural gas (LNG) purchases, forming strategic alliances with Namibia, Egypt and Kazakhstan to guarantee a stable and sustainable supply of renewable hydrogen, and concluding agreements for the export of natural gas to Europe with Israel and Egypt (European Commission, 2022).

In order to reduce its reliance on Russian gas, the EU focused on securing affordable energy supplies, saving energy, and investing in renewables. In 2022, renewable energy accounted for 23% of the EU's total energy consumption (Eurostat, 2023). Sweden led among the EU countries, as 66.0% of its energy consumption came from renewable sources in 2022. In the EU, the percentage of gross electricity consumption that came from renewable sources was 41.2% in 2022 (Eurostat, 2023). The EU's capacity to produce renewable energy from solar and wind power surpassed 400 GW in 2022 (European Commission, 2023a). The investment in renewable energy remains a priority for supporting clean transition.

Spikes in energy costs, post-COVID supply chains disruptions labor shortages, and high inflation posed serious challenges to the industry (European Commission, 2023a). The EU also unveiled the Green Deal Industrial Plan, which aims to navigate the shift to carbon neutrality. The goal of the Green Deal Industrial Plan is to simplify, expedite, and harmonize incentives in order to maintain the EU's competitiveness and allure as an investment destination for the net-zero industry (European Commission, 2023a). The Plan has four main pillars: predictable and simplified regulatory environment, faster access to funding, enhancing skills, and open trade for resilient supply chains (European Commission, 2023a). The Plan seeks to establish a more straightforward and predictable system that would guarantee that consumers would profit from renewable energy sources' low costs. The EU's commitment to accelerating the domestic shift to a net-zero economy is emphasized in the Green Deal Industrial Plan. It is underlined that in addition to meeting demands and seizing opportunities like expanding the use of renewable energy sources, modernizing energy and transportation infrastructure like grids, and dramatically shifting to fossil-free hydrogen as a fuel, feedstock, and storage medium, the EU can also take the lead in developing the future net-zero industries (European Commission, 2023a, p. 2). In 2019, the number of green jobs was 4.5 million, indicating an increase compared to 2000s (European Commission, 2023b). In 2021, the value of the EU's net-zero start-up ecosystem was over EUR 100 billion. The International Energy Agency projects that by 2030, the global market for key mass-produced renewable energy technologies would worth approximately USD 650 billion annually (European Commission, 2023b).

In relation to the Green Deal Industrial Plan, President of the European Commission, von der Leyen, stated that the EU has a chance to lead the way in the rapidly expanding net-zero technology sector with a clear purpose. She highlighted that Europe is committed to spearheading the shift in clean technology (European Commission, 2019d). The plan is significant as its main objective is to chart a clear path for the industry and people throughout the transition. The Green Deal Industrial Plan emphasized that for the green transition to provide fair and just outcomes, it must be inclusive and people-centered.

## Objectives and Commitments of the European Union for Climate Action

### *Reducing Greenhouse Gas Emissions*

The 2030 Agenda for Sustainable Development, which was approved by the UN in 2015, is a comprehensive action plan for the prosperity of people and the planet. The Agenda includes 17 SDGs and 169 objectives with focus on economic, social, and environmental aspects, which are three factors for sustainable development, relying on the previous eight Millennium Development Goals (United Nations, 2015a). As the European Commission rolled out the EGD in 2019 to attach priority to sustainability and intended to become climate neutral by 2050, the prospective policymaking attempts echo an innovation-based agenda in line with the UN SDGs. The current challenge facing the EU is ensuring that the agenda is implemented effectively (McCann & Soete, 2020).

The EU has made a great deal of attempts that indicate its commitment to bioremediation and sustainable transformation. Among them are the EGD and NextGenerationEU recovery plan, which play a crucial role in promoting environmentally friendly policies and sustainable practices across the bloc (Carnazza et al., 2023).

Adopted by the EU as a postpandemic recovery plan, the NextGenerationEU adds to the EGD implementation and deposits no less than 37% of the Recovery and Resilience Facility (RRF) into climate actions for the EU member states (European Commission, 2021). As actions for green transformation are adopted, an emphasis is placed on consideration of distribution effects by the member states, the adaptation to the European Social Charter principles, and how the Recovery and Resilience Plan would achieve a fair transition. These policy instruments point to a gradual development of an EU framework that combines social and environmental policies and to the various forms of developing climate action policies for the EU. However, winning popular approval is crucial for Europe to politically implement the green transformation (Baute, 2024).

Given the current state of the EU, all large-scale facilities are governed under an EU-wide classification as per the EU ETS, and most of the greenhouse gas emissions are tackled by the EU member states in line with the individual and binding emission reduction goals. The EU has adopted the regulatory nature of the Kyoto Protocol that comes with individual emission reduction commitments with a variety of emissions trading options available for the EU member states (Peeters & Athanasiadou, 2020). The triple sectoral regime was originally based on the implementation of different rules for energy-intensive industries and domestic market sectors. A per-capita carbon emissions regime has been adopted by the domestic market sectors.

The Effort Sharing Decision and the Effort Sharing Regulation, which were implemented in 2009 and 2018, respectively, substantially improved the regulatory effort sharing framework for non-ETS emissions. In terms of the criteria adopted to set different national goals, both rely on the financial performance, that is to say the capacity measured by the gross domestic product (GDP) per capita. In 2021, a goal was set to reduce emissions by 55% by 2030 in comparison to 1990 (Steininger et al., 2022).

The EU Effort Sharing Regulation sets a national goal on how to reduce the greenhouse gas emissions by 2030 for each EU member state. The goal also includes emissions caused by sectors including nonaviation domestic transportation, buildings, agriculture, small-sized industries, and waste. The share of EU domestic emissions covered by the Effort Sharing Regulation is about 60%. The Regulation adopted in 2018 was revised in 2023. The revision was adopted as a part of a series of recommendations intended to reduce the EU's emissions by 55% by 2030 (European Commission, 2024b).

As shown in Table 13.1, the EU member states now intend to reduce emissions by 10%–50% compared to 2005.

Table 13.1. Emission Reduction Goals of the Member States by 2030.

| Country | Old Targets (set in 2018) | New Targets (since 2023) |
|---|---|---|
| Belgium | −35% | −47% |
| Bulgaria | 0% | −10% |
| Czechia | −14% | −26% |
| Denmark | −39% | −50% |
| Germany | −38% | −50% |
| Estonia | −13% | −24% |
| Ireland | −30% | −42% |
| Greece | −16% | −22.70% |
| Spain | −26% | −37.70% |
| France | −37% | −47.50% |
| Croatia | −7% | −16.70% |
| Italy | −33% | −43.70% |
| Cyprus | −24% | −32% |
| Latvia | −6% | −17% |
| Lithuania | −9% | −21% |
| Luxembourg | −40% | −50% |
| Hungary | −7% | −18.70% |
| Malta | −19% | −19% |
| Netherlands | −36% | −48% |
| Austria | −36% | −48% |
| Poland | −7% | −17.70% |
| Portugal | −17% | −28.70% |
| Romania | −2% | −12.70% |
| Slovenia | −15% | −27% |
| Slovakia | −12% | −22.70% |
| Finland | −39% | −50% |
| Sweden | −40% | −50% |

Source: European Commission, 2024b.

Sharing out the burdens of achieving the 2030 goals to reduce carbon emissions must be in harmony with the achievement of adequate prosperity for all nationals of the member states. The principle of equality is a testament to the principle of adequacy in this sense (Steininger et al., 2022). Under the principle of accountability, the states must be held accountable for their own greenhouse gas emissions because they are informed of the capability to absorb the limited capacity of the atmosphere. The states must acknowledge their share in using this limited resource and recognize that most of the regimes related to sharing out the remaining carbon budget require categorically reducing the emissions.

The Just Transition Fund is one of the key instruments to promote the EU member states in achieving climate neutrality by 2050. In 2019, the European Commission recommended creating a Just Transition Mechanism including a Just Transition Fund for the effective and just achievement of climate neutrality in the EU as a part of the EGD that sets a roadmap for the EU's new policy of growth. It highlighted the fact that the Just Transition Mechanism needs to focus on territories and sectors hit the hardest by transition as they rely on fossil fuels and greenhouse-intensive industrial processes (European Commission, 2022). The Just Transition Fund is a dedicated program created as a part of the InvestEU program and intended to mobilize additional investments in relevant territories through public sector loans provided by the European Investment Bank. The Fund initially provides a grant, while the territorial InvestEU transition plan focuses on private investments. Public funds must be used for the actions of the European Investment Bank. The Just Transition Fund aims to enhance economic diversity in regions most affected by climate change, support the retraining of job seekers, and help them actively participate in the workforce (European Commission, 2022).

The European Commission pursues the goal of "becoming the world's first climate neutral continent" by 2050 and growing into a global standard setter. EGD is advocated as a policy agenda that helps European businesses achieve innovation, develop new technologies, create new markets, and take advantage of the competitive edge and equal conditions. The EU, therefore, intends to become the leader of the global regulatory competition and serve as a role model, disseminating the environmental norms. However, EGD has an influential inward component intended to safeguard the environmental assets and economic interests of Europe as a new growth and industrial strategy. These latter aspects are more compatible with the geopolitical and geo-economic role of the EU, which has been hotly debated in recent times (Eckert, 2021).

Reducing gas emissions sits at the heart of the EGD, and individual actors within the EU are expected to take action to reach the shared goal of climate neutrality set by the EU. The EU has taken decisive actions to achieve the goals set for climate action. The greenhouse gas emission reduction goals adopted by the EU are a major commitment for sustainable development. In addition, the EU seeks to conform to the principles of being just and equal, providing support to the territories hit the hardest by climate change through financial instruments including the Just Transition Fund. The steps taken by the EU serve as a model for domestic and global climate actions and lend a momentum to global efforts for sustainability.

### *Circular Economy*

The global population is estimated to hit 9.4 billion by 2050, and the growing population is expected to cause a major boost in consumption demands. Furthermore, such large-scale production and consumption will give rise to excessive use of natural resources and growing amounts of waste. This means that the linear economy falls short in terms of sustainability (Song et al., 2019). The current economic systems need to be revised in consideration of sustainability, and use of resources and other options including the circular economy need to be entertained.

In the aftermath of the decades that followed World War II, challenges arose with regards to the linear economy as the regeneration capacity of the planet dwindled. At the time, people recognized how the linear economy regime fell short in terms of sustainability and looked for other options. As a result of a growing demand for goods, the growing supply causes challenges in access to resources and growing amount of waste in a linear economy. This gives rise to major environmental, social, and financial harms. That is why the circular economy looms large as a system that causes less environmental harm compared to the linear economy. The conventional linear economy is built on constant growth of mass production and consumption. However, this regime causes not only the depletion of natural resources but also the growing pile of waste worth almost zero (Ilinova & Kuznetsova, 2022). This leads to an unsustainable situation in economic, environmental, and social aspects.

The circular economy has been described in various ways, and it is still a burgeoning field of interest (Kirchherr et al., 2017). The circular economy stands for establishing a variety of circular processes by linking goods with the fact that they must have their own circle, maximizing the use of goods content, reintegrating goods into the production process, keeping the environmental impact of the process at minimum, and safeguarding the concept of repair. Given that goods need to have their own circle, the circular economy involves the description of various processes of circularity, optimal use of goods content, and reintegration into the production process. It is intended to keep the environmental impact of the process at minimum and safeguard the concept of repair (Ellen MacArthur Foundation, 2013). The circular economy is a concept that restricts the use of resources and embodies a mindset that attaches importance to repair (Ellen MacArthur Foundation, 2013). While the production process typically consists of take-make-use phases, the circular economy stands for reintegrating goods, which are not used and have yet to expire, back into production, and thus building a circle that reduces the use of resources (Ripanti et al., 2015).

The goals of the circular economy are as follows: reusing waste in a production process, building a resilient system through diversity as a part of the circular economy, taking actions to use energy from renewable resources, keeping production and consumption in a closed circle, and making a gradual plan for transition to circular economy (Ellen MacArthur Foundation, 2013).

The growing population and the boom in use of resources pose a threat to the natural equilibrium of the global ecosystem. That is why the transition to circular economy rather than linear economy is a must for a sustainable future (Lehner et al., 2016). The growing population and the boom in use of resources pose a threat to sustainability, pushing the limits of natural resources. Adopting a circular economy may mitigate the threat, helping make more efficient and effective use of the resources.

Adopting the concept of circular economy reduces the excessive use of the resources and mitigates the effects of excessive consumption, adding to sustainability. This approach helps reduce the consumption of resources, cope with the lack of resources and climate change, and is likely to mitigate the carbon emission and use of raw materials. This may improve the efficacy of climate action (Polyportis et al., 2022; Tunn et al., 2019).

Adopted in 2015, the first CEAP involves 54 actions, with an intention to convert the European economy from a linear regime to a circular one. All of 54 actions were adopted or implemented by 2019 (European Environment Agency, 2022). The circular economy activities grew by 6% from 2012 to 2016 in the EU (Eurostat, 2022).

In an effort to make efficient use of resources and achieve carbon neutrality as a part of the EGD, the EU rolled out a new economic regime, adopting the New CEAP. The New CEAP is intended to "maximize the lifespan of goods, protect the environment, safeguard consumer rights, and improve the competitive edge of producers. A policy fit for a circular economy is adopted to conserve the resources to the extent possible for a sustainable future" (TÜSİAD, 2021).

The objectives of the New CEAP are as follows (European Commission, 2024a):

- Creating sustainable goods in the EU.
- Empowering the household and the public.
- Attaching priority to resource-intensive sectors for transition to circular economy (electronics, batteries-vehicles, packaging, plastics, textile, construction-buildings, food, water, and food stuff, etc.).
- Focusing on resource-intensive sectors where there is more potential for circularity.
- Waste reduction.
- Circularity being an operational system for people, cities, and territories.
- Leading global efforts for circular economy.

Circular economy rediscovers the European economy and makes it sustainable and competitive. Benefits of the circular economy for businesses, industries, and people in Europe are listed as follows (Eurostat, 2022):

- Having access to more innovative and efficient means of production and consumption.
- Protecting businesses from lack of resources and volatility in prices.
- Offering opportunities for local jobs and social integration.
- Optimizing waste management that expedites recycling and reduces the landfills.
- Saving energy since less production requires less energy.
- Climate and biodiversity help protect the environment from air, soil, and water pollution.

The European Commission offers a comprehensive perspective in its action plans to promote the circular economy. Table 13.2 shows the actions taken and attempts made by the EU for transition to the circular economy:

Table 13.2. Timeline of Actions Taken During the Transition to a Circular Economy.

| Year | Actions |
|---|---|
| December 2015 | The European Commission adopted its first CEAP. |
| December 11, 2019 | The European Commission approved the EGD. |
| March 11, 2020 | The European Commission introduced its new CEAP. |
| December 10, 2020 | The European Commission approved a proposal for a new regulation on sustainable batteries. |
| February 22, 2021 | The Global Alliance on Circular Economy and Resource Efficiency (GACERE) was launched. |
| October 28, 2021 | The European Commission approved a proposal to update the rules on persistent organic pollutants in waste. |
| November 17, 2021 | The European Commission approved a proposal including new rules on shipments of waste. |
| March 30, 2022 | The European Commission adopted a package of measures proposed in the CEAP.<br>Regulation on Sustainable Products<br>EU's strategy for sustainable and circular textiles<br>Proposal for a revised Building Materials Regulation<br>Proposal to empower consumers in green transformation |
| April 5, 2022 | The European Commission approved revised proposals for EU measures to address pollution from large industrial installations.<br>Revision of the Directive on Industrial Emissions<br>Revision of the European Pollutant Release and Transfer Register |
| 2022 | Within the scope of the Circular Action Plan, the following initiatives were launched:<br>Legislative proposal to verify green claims made by companies<br>Review of packaging and packaging waste requirements in the EU |

*(Continued)*

Table 13.2. *(Continued)*

| **Year** | **Actions** |
|---|---|
| | Establishing a new policy framework for bio-based, biodegradable, and compostable plastics<br>Taking measures to reduce the impact of microplastic pollution on the environment |
| March 22, 2023 | The European Commission approved a set of proposals on green claims and adopted the right to repair.<br>Proposal for the Green Claims Directive<br>Proposal on common rules promoting the repair of goods |
| May, 2023 | The European Commission revised its framework for monitoring the circular economy. |
| September–October 2023 | Several initiatives related to microplastics were launched:<br>The Reach Re striction of intentionally added microplastics<br>Proposal for a Regulation on preventing pellet losses to reduce microplastic pollution<br>Brochure on EU action against microplastic pollution |

Source: European Commission, 2024a.

The new Circular Action Plan introduced specific measures with the following aims (European Commission, 2024a):

- To make sustainable products the norm in the EU.
- Empower consumers and public buyers.
- Focus on the sectors that use the most resources and where the potential for circularity is high such as electronics and ICT, batteries and vehicles, packaging, plastics, textiles, construction and buildings, food, water, and nutrients.
- Reduce waste.
- Make circularity work for people, regions, and cities.
- Lead global efforts on circular economy.

Various methods are used to monitor member states' transition to the circular economy. The Bellagio Declaration was developed to monitor the progress and guide institutions (European Environment Agency, 2022). This declaration set the basic principles of the transition to a circular economy to allow the process to be monitored. However, the European Commission and Eurostat also developed the European Circular Economy Monitoring Framework to monitor the progress toward a circular economy (European Environment Agency, 2022). This monitoring

Table 13.3. EU's Key Indicators for Monitoring Circular Economy.

| **Production and Consumption** | **Material Consumption Waste Generation** |
|---|---|
| Waste Management | Overall recycling rates<br>Recycling rates for specific waste streams |
| Secondary raw materials | Contribution of recycled materials to raw material demand<br>Recyclable raw materials trade |
| Competitiveness and innovation | Private investment, employment, and gross added value related to circular economy sectors<br>Innovation |
| Global sustainability and resilience | Global sustainability from the circular economy<br>Resilience from the circular economy |

Source: Eurostat, 2024.

framework, developed by the European Commission, includes five key indicators divided into specific categories and provides key factors of the circular economy. Table 13.3 shows five key indicators along with their description.

The transition to a circular economy, which is at the heart of the EGD, is expected to reduce pressure on natural resources and create sustainable growth and more jobs. It will also help the EU achieve its 2050 climate neutrality target and halt biodiversity loss (European Commission, 2024a).

## The Global Importance of EU's Climate Policy

The EU is recognized as a strategic international player that can act as a community in cooperation with its neighbors and global strategic partners, based on the norms and values of stability and cooperation. However, it is necessary to scrutinize whether this is adequately reflected in the EU's current external role, as a strategic power that has the capacity to meet natural and legitimate expectations on issues such as environmental governance (Xavier, 2019).

Four conditions are to be met for the European Union to be considered a truly global player (Jupille & Caporaso, 1998):

- Recognition: Recognition and interaction with other countries.
- Authority: Legal competence of the EU on a specific subject.
- Autonomy: Showing a degree of institutional difference and independence from other actors.
- Harmony: The degree to which an entity formulates and articulates internally consistent policy preferences.

The set of conditions highlight that the policies set by the EU need to be realized in harmony which also necessitates commitment to the objectives. Thus, the Union can act as a key actor which has the capacity to project its power to international relations and shape global policies.

The EU's policies regarding green transition and climate neutrality have drawn global attention as the EGD was even described as the "next Industrial Revolution." The climate change's devastating impact urges states to act and transform their economies. Therefore, the EU's policies and their implementation regarding the EGD are closely viewed by the non-EU actors.

International cooperation is essential to address the issue of climate change. The COP and G20 lead advances in global cooperation. The International Monetary Fund (IMF) has already included issues related to climate change in its analyses, recommendations, and policies. In this context, the EU plays a vital role in its dedication and commitment to reducing greenhouse gas emissions. The CBAM, launched by the EU, aims to prevent carbon leakage to other regions. It will be supported by an agreement on the international carbon price floor mechanism, which is currently being discussed at international platforms and organizations, including the IMF. The EU is also in the process of developing new environmental, social, and governance (ESG) commitments while reiterating its commitment to the climate transition goal in middle- and low-income countries. All of these efforts can contribute to consolidating and maintaining the EU's leadership in the global climate transition process (Dormido et al., 2023).

The EU has long been pursuing its goal to lead global efforts on climate policy and play a trailblazer role in efforts to reduce carbon emissions (Rayner et al., 2023). These efforts have not only contributed to the shaping of international institutional regulations but have also ensured the development of policies and measures to reduce greenhouse gas emissions. Since the 1990s, climate-related policy efforts had major impacts within and beyond the EU's borders and have been covered by expanding economic sectors over time. The fight against climate change has also been incorporated into the EU law. Although the EU's founding agreements do not directly define a climate policy, Article 191(1) of the Treaty on the Functioning of the European Union lists combating climate change as one of its main objectives within the scope of environmental protection policy (Stoczkiewicz, 2018). In 2019, the European Commission introduced a set of measures within the scope of the EGD and brought climate change to the top of the EU's legislative agenda. The package of measures also included the EU's goal of reducing the net carbon emission level to zero within half a century (European Commission, 2019b).

The EU has set an example for other regions by adopting new policy instruments such as the ETS. Although the EU's financial resources are limited compared to its member states, it has made significant contributions and allocated an increasing proportion of the EU budget toward climate goals (Rietig, 2021). Furthermore, the EU members tend to accept more ambitious emission reduction targets as well as increased climate resilience goals (Biesbroek & Swart, 2019). As recognized by the members, climate change is not only a social problem but also an economic problem (Pandya & Leal-Arcas, 2024).

In recent years, international actors have placed a greater focus on combating climate change and ensuring sustainable growth, as one of the most important issues on the international agenda. To reduce pollutant emissions, several countries must adopt effective mitigation measures based on international agreements. International cooperation plays a pivotal role in terms of fulfilling the commitments made under these agreements, ensuring the energy transition, and halting climate change. Countries with the largest greenhouse gas emissions, such as some G20 countries, and the IMF started to show greater concern about climate issues while fulfilling their duties. The EU remains an active and leading player in terms of this global commitment, setting increasingly ambitious goals. The EU emphasized the climate neutrality target in the European Climate Law in accordance with the EGD and initiated several groundbreaking policies such as Fit for 55 (European Parliament, 2023) to achieve this goal (Dormido et al., 2023).

The EU's climate diplomacy agenda is driven by the need for coordinated expansion of climate action between different EU institutions and member states. The annual Climate Diplomacy Week brings various actors together to set a common agenda and provide joint social support. During this week, EU Delegations and embassies of EU member states around the world organize events and share good practices to foster cooperation and dialog on climate change, while calling for more action in this field (European Parliament, 2021).

Key tasks of the EU Climate Diplomacy for the future include ensuring that public and private investment flows are compatible with the climate, promoting a just transition both domestically and internationally during the transition to a green economy, effectively addressing and managing the geopolitical challenges that arise in the process of energy transformation (European Parliament, 2021).

The invasion of Ukraine by Russia in 2022 has led to changes in the EU's climate strategy while accelerating the implementation of initial plans for the transformation of the energy model. Given the importance of Russia's oil and, above all, gas supply to the EU, member states aim to decarbonize energy, reduce the EU's dependence on gas, and achieve a model consistent with its impact on the setting of energy prices in the EU and its EGD objectives. However, disruptions and increasing costs in European and global energy markets due to the war may also make it economically and socially difficult to achieve these goals (Dormido et al., 2023).

Overall, the EU has adopted one of the highest environmental standards worldwide, leading to a greener economy. This approach aims to protect the health and life quality of people living in the EU by protecting natural resources. In addition to this, the EU also seeks to create a more competitive economy that promotes employment and increases investments. It balances this approach with a sustainable environmental framework focusing primarily on improving people's quality of life. However, despite the principle of delegation, the EU institutions still assume the regulatory role (Russo et al., 2022; Xavier, 2019)

As a global actor, the EU led environmental efforts to achieve sustainable development and green economy goals. It also aims to improve human health by ensuring higher environmental standards and protecting nature. While the environmental policies of the EU are harmonized among the member states, it seeks to increase its authority on the international stage. More action is needed to meet the needs of future generations (Xavier, 2019). The EU's focus on sustainable development and green economy plays a critical role in terms of environmental governance and strengthening international norms. The devastating impact of the climate change adds pressure on states to take urgent action. The EU's policies on climate neutrality and their implementation might be viewed as example by non-EU actors as the fight against climate change is included in the agenda of world politics.

## Conclusion

Global action is a sine qua non for mitigating unprecedented impact of the climate change. As experts and scientists call for urgent action, the international actors feel the pressure to address climate crisis on their political agenda. The EU introduced the ambitious Green Deal to take decisive action to reverse the trends of climate change.

The EGD sets out the vision of Europe becoming the first climate-neutral continent by 2050. With comprehensive set of laws and mechanisms, the EGD also urges member countries to translate climate objectives into realities. While working on the transition to a net-zero economy, the world and Europe has faced challenges such as the COVID-19 pandemic, the war in Ukraine, and the consequences of being dependent on imported gas. These challenges prompted Europe to enhance efforts for creating secure and sustainable ways to ensure the well-being its members. REPowerEU puts an emphasis on energy savings and renewable energy deployment to sustain energy security of Europe. It is estimated that if EU laws are implemented effectively and on time by 2030, 55% of electricity will come from wind and solar sources. Compared to 2019, gas consumption will be reduced by at least 31% and oil consumption by 34%. Additionally, there will be 29 million private electric cars on European roads (Makaroff & Kalcher, 2022).

The Europe's Green Deal indicates the political will to take climate action and undergo a comprehensive transformation encompassing various sectors of the economy. The EGD highlights the EU's efforts to build a new energy security architecture and ecosystem that would not only navigate the climate action of member states but also set an example for other international actors. Fit for 55 and REPowerEU set the foundation for this new energy security architecture drawing up strategies to energy transition. However, the Green Deal has been also facing opposition from various sectors across Europe. The EU dropped a reference to the possibility for agriculture to reduce non-CO2 emissions by at least 30% in 25 years in its 2040 climate roadmap while protests of the farmers were ongoing in Europe against the EU's green policies (Arboleas, 2024). Implementing of policies and regulations under the EGD might be considered costly by some sectors. To reach the objectives of the ambitious Green Deal, all members are expected to be on the same page and show the will to sustain the transition to the climate neutrality, paying heed to the needs of the key sectors. Regarding the recommendation of 90% net emissions reduction by 2040 issued on February 6, 2024, it is highlighted that the EU needs to "capture at least 50 million tonnes of CO2 per year by 2030, approximately 280 million tonnes by 2040, and around 450 million tonnes by 2050" (European Commission, 2024c).

The EU's climate diplomacy and action plans aim to ensure that the EU has a strong voice in the fight against climate change on the international stage. In particular, the EU's policies and programs, such as reducing carbon emissions, energy transformation, and investing in green technology, contribute to the acceleration of climate action on a global scale. However, further cooperation and partnership development efforts are needed to increase the EU's impact on the international arena. Accordingly, it can be said that the EU's climate policies reflect its commitment to generating an effective solution to the climate change crisis with the joint efforts of the international community. The EGD presents a holistic approach to navigate economies facing unprecedented climate challenges. For achieving the goals for a sustainable future, commitment to the climate agenda remains essential.

## References


Arboleas, M. S. (2024, Şubat 6). EU Commission backtracks on agricultural emissions cuts. https://www.euractiv.com/section/agriculture-food/news/eu-commission-backtracks-on-agricultural-emissions-cuts/

Baute, S. (2024). The distributive politics of the green transition: A conjoint experiment on EU climate change mitigation policy. *Journal of European Public Policy*, 1–29.

Biesbroek, R., & Swart, R. (2019). Adaptation policy at supranational level? Evidence from the European Union. *Research Handbook on Climate Change Adaptation Policy*, 194–211.

Carnazza, G., Renström, T. I., & Spataro, L. (2023). *Is public debt environmentally friendly? The role of EU fiscal rules on environmental quality: An empirical assessment*.

Dormido, L., Garrido, I., L'Hotellerie-Fallois, P., & Santillán, J. (2023). Climate change and sustainable growth: International initiatives and European policies. There Is a Spanish Version of This Edition with the Same Number.

Eckert, S. (2021). The European Green Deal and the EU's regulatory power in times of crisis. *Journal of Common Market Studies*, *59*, 81–91.

Ellen MacArthur Foundation. (2013). *Towards the Circular Economy: Economic and business rationale for an accelerated transition*.

EUR-Lex. (2020, March 11). *Communication from the Commission to the European Parliament, the Council, the European Economic and Social Committee and the Committee of the Regions*. https://eur-lex.europa.eu/legal-content/EN/TXT/?qid=1583933814386&uri=COM:2020:98:FIN

European Commission. (2019a). A European Green Deal: Striving to be the first climate-neutral continent. https://ec.europa.eu/info/strategy/priorities-2019-2024/european-green-deal_en

European Commission. (2019b). What is the European Green Deal? https://ec.europa.eu/commission/presscorner/api/files/attachment/859152/What_is_the_European_Green_Deal_en.pdf.pdf
European Commission. (2019c). Climate action and the Green Deal. https://commission.europa.eu/strategy-and-policy/priorities-2019-2024/european-green-deal/climate-action-and-green-deal_en
European Commission. (2019d). The European Green Deal. https://commission.europa.eu/strategy-and-policy/priorities-2019-2024/european-green-deal_en
European Commission. (2021). Guidance to member states recovery and resilience plans. SWD. PART 1/2.
European Commission. (2022). Just transition fund. https://www.europarl.europa.eu/factsheets/en/sheet/214/just-transition-fund-jtf
European Commission. (2023a, February 1). Communication from the Commission to the European. https://commission.europa.eu/document/download/41514677-9598-4d89-a572-abe21cb037f4_en?filename=COM_2023_62_2_EN_ACT_A%20Green%20Deal%20Industrial%20Plan%20for%20the%20Net-Zero%20Age.pdf
European Commission. (2023b, February 1). Factsheet: The Green Deal industrial plan. https://ec.europa.eu/commission/presscorner/detail/en/FS_23_514
European Commission. (2023c). Carbon border Adjustment mechanism. https://taxation-customs.ec.europa.eu/carbon-border-adjustment-mechanism_en
European Commission. (2023d). Leading the clean tech revolution. *Retrieved from European Commission*. https://commission.europa.eu/strategy-and-policy/priorities-2019-2024/european-green-deal/green-deal-industrial-plan_en
European Commission. (2024a). Circular economy action plan. https://environment.ec.europa.eu/strategy/circular-economy-action-plan_en
European Commission. (2024b). Effort sharing 2021-2030: Targets and flexibilities. https://climate.ec.europa.eu/eu-action/effort-sharing-member-states-emission-targets/effort-sharing-2021-2030-targets-and-flexibilities_en
European Commission. (2024c, February 6). *Commission sets out how to sustainably capture, store and use carbon to reach climate neutrality by 2050*. https://ec.europa.eu/commission/presscorner/detail/en/ip_24_585
European Commission. (2024d, January 31). *Carbon border Adjustment mechanism (CBAM) Questions and Answers*. Retrieved from European Commission. https://taxation-customs.ec.europa.eu/system/files/2023-12/Questions%20and%20Answers_Carbon%20Border%20Adjustment%20Mechanism%20%28CBAM%29.pdf
European Commission. (2024e). 2040 climate target. https://climate.ec.europa.eu/eu-action/climate-strategies-targets/2040-climate-target_en
European Commission. (2020a). Biodiversity strategy for 2030. https://environment.ec.europa.eu/strategy/biodiversity-strategy-2030_en
European Commission. (2021b). European climate law. https://climate.ec.europa.eu/eu-action/european-climate-law_en
European Commission. (2021c). Fit for 55: Delivering on the proposals. https://commission.europa.eu/strategy-and-policy/priorities-2019-2024/european-green-deal/delivering-european-green-deal/fit-55-delivering-proposals_en
European Commission. (2020b). On the path to a climate-neutral Europe by 2050. https://commission.europa.eu/strategy-and-policy/priorities-2019-2024/european-green-deal/delivering-european-green-deal_en

European Commission. (2018). Renewable energy targets. https://energy.ec.europa.eu/topics/renewable-energy/renewable-energy-directive-targets-and-rules/renewable-energy-targets_en

European Commission. (2022). REPowerEU at a glance. https://commission.europa.eu/strategy-and-policy/priorities-2019-2024/european-green-deal/repowereu-affordable-secure-and-sustainable-energy-europe_en

European Council. (2019e). Paris Agreement on climate change. https://www.consilium.europa.eu/en/policies/climate-change/paris-agreement/#:~:text=The%20EU%20and%20the%20Paris%20Agreement,-Infographic%20%2D%20Paris%20Agreement&text=As%20required%20by%20the%20agreement,2030%2C%20compared%20to%201990%20levels

European Environment Agency. (2022). Measuring Europe's circular economy. https://www.eea.europa.eu/themes/waste/measuring-europes-circular-economy

European Parliament. (2021). EU climate change diplomacy in a post-Covid-19 world. https://south.euneighbours.eu/wp-content/uploads/2022/07/QA0221895ENN.en_-1.pdf

European Parliament. (2023). Social climate fund: 'Fit for 55' package. *Briefing: EU Legislation in progress.* https://www.europarl.europa.eu/RegData/etudes/BRIE/2021/698777/EPRS_BRI(2021

Eurostat. (2022). Overview circular economy. https://ec.europa.eu/eurostat/web/circular-economy

Eurostat. (2023, December). *Renewable energy statistics.* https://ec.europa.eu/eurostat/statistics-explained/index.php?title=Renewable_energy_statistics#Wind_and_water_provide_most_renewable_electricity.3B_solar_is_the_fastest-growing_energy_source

Eurostat. (2023, December 22). *23% of energy consumed in 2022 came from renewables.* https://ec.europa.eu/eurostat/web/products-eurostat-news/w/ddn-20231222-2#:~:text=The%20share%20of%20renewable%20sources,1.1%20percentage%20points%20(pp)

Eurostat. (2024). *Circular economy monitoring framework.* https://ec.europa.eu/eurostat/web/circular-economy/monitoring-framework

Ilinova, A., & Kuznetsova, E. (2022). CC(U)S initiatives: Prospects and economic efficiency in a circular economy. *Energy Reports*, *8*, 1295–1301. https://doi.org/10.1016/J.EGYR.2021.11.243

Jupille, J., & Caporaso, J. A. (1998). *States, agency, and rules: The European Union in global environmental politics* (Vol. 17, pp. 157–182). The European Union in the World Community.

Kirchherr, J., Reike, D., & Hekkert, M. (2017). Conceptualizing the circular economy: An analysis of 114 definitions. *Resources, Conservation and Recycling*, *127*, 221–232. https://doi.org/10.1016/J.RESCONREC.2017.09.005

Lehner, M., Mont, O., & Heiskanen, E. (2016). Nudging–A promising tool for sustainable consumption behaviour? *Journal of Cleaner Production*, *134*, 166–177.

McCann, P., & Soete, L. (2020). *Place-based innovation for sustainability*. Publications Office of the European Union.

Niranjan, A. (2023, July 27). 'Era of global boiling has arrived,' says UN chief as July set to be hottest month on record. https://www.theguardian.com/science/2023/jul/27/scientists-july-world-hottest-month-record-climate-temperatures

Pandya, D., & Leal-Arcas, R. (2024). The creation of a climate club for a sustainable economic future: The role of international economic law amidst geopolitical confrontation. *Georgia Journal of International & Comparative Law*, *52*(1), 1–88.

Peeters, M., & Athanasiadou, N. (2020). The continued effort sharing approach in EU climate law: Binding targets, challenging enforcement? *Review of European, Comparative & International Environmental Law*, *29*(2), 201–211.

Polyportis, A., Mugge, R., & Magnier, L. (2022). Consumer acceptance of products made from recycled materials: A scoping review. *Resources, Conservation and Recycling*, *186*, 106533. https://doi.org/10.1016/J.RESCONREC.2022.106533

Rayner, T., Szulecki, K., Jordan, A. J., & Oberthür, S. (2023). The global importance of EU climate policy. In *Handbook on European Union climate change policy and politics* (pp. 1–21). Edward Elgar Publishing.

Rietig, K. (2021). Accelerating low carbon transitions via budgetary processes? EU climate governance in times of crisis. *Journal of European Public Policy*, *28*(7), 1018–1037.

Ripanti, E. F., Tjahjono, B., & Fan, I.-S. (2015). Circular economy in reverse logistics: Relationships and potential applications in product remanufacturing circular PSS supply chains: How to organize supply and delivery chains for circularity view project health management of RPAS view project. https://www.researchgate.net/publication/282007254

Russo, V., Bilucaglia, M., Circi, R., Bellati, M., Valesi, R., Laureanti, R., Licitra, G., & Zito, M. (2022). The role of the emotional sequence in the communication of the territorial cheeses: A neuromarketing approach. *Foods*, *11*, 2349. https://doi.org/10.3390/foods11152349

Song, L., Lim, Y., Chang, P., Guo, Y., Zhang, M., Wang, X., Yu, X., Lehto, M. R., & Cai, H. (2019). Ecolabel's role in informing sustainable consumption: A naturalistic decision making study using eye tracking glasses. *Journal of Cleaner Production*, *218*, 685–695. https://doi.org/10.1016/j.jclepro.2019.01.283

Steininger, K. W., Williges, K., Meyer, L. H., Maczek, F., & Riahi, K. (2022). Sharing the effort of the European Green Deal among countries. *Nature Communications*, *13*(1), 3673.

Stoczkiewicz, M. (2018). The climate policy of the European Union from the framework convention to the Paris agreement. *Journal for European Environmental & Planning Law*, *15*(1), 42–68.

Strategic Perspectives. (2022). Turning the European deal into reality. https://strategicperspectives.eu/wp-content/uploads/2023/05/StrategicPerspectives_TurningtheEuropeanGreenDealintoReality_May2023-2.pdf

The World Bank. (2023, May 12). *Climate change*. https://www.worldbank.org/en/topic/climatechange/overview#:~:text=According%20to%20our%20research%2C%20climate,world's%20most%20food%2Dinsecure%20regions

Tunn, V. S. C., Bocken, N. M. P., van den Hende, E. A., & Schoormans, J. P. L. (2019). Business models for sustainable consumption in the circular economy: An expert study. *Journal of Cleaner Production*, *212*, 324–333. https://doi.org/10.1016/J.JCLEPRO.2018.11.290

TÜSİAD. (2021). *Türkiye'de Döngüsel Ekonomiye Geçiş Tutum Belgesi*. https://www.tusiad.org/tr/haberler/item/10832-turkiye-de-dongusel-ekonomiye-gecis-tutum-belgesi

United Nations. (2015a). *Transforming our world: The 2030 agenda for sustainable development.*

United Nations. (2015b). *Key aspects of the Paris agreement.* Retrieved from United Nations. https://unfccc.int/most-requested/key-aspects-of-the-paris-agreement#:~:text=The%20Paris%20Agreement%E2%80%99s%20central%20aim,further%20to%201.5%20degrees%20Celsius

Watts, J. (2018, March 19). Water shortages could affect 5bn people by 2050, UN report warns. https://www.theguardian.com/environment/2018/mar/19/water-shortages-could-affect-5bn-people-by-2050-un-report-warns

Xavier, A. I. (2019). The European union as a leading environmental player? A critical analysis on the policy and commitments towards global development and climate change. *Climate Change and Global Development: Market, Global Players and Empirical Evidence*, 135–1.

Chapter 14

# Türkiye's Strides Toward a Greener Future: National and International Political Progress on Climate Change

*Büşra Deniz*[1], *Derya Ulutürk*[2] *and Mustafa Başkara*[3]

Ankara University National Center for the Sea and Maritime Law (DEHUKAM), Türkiye

## Abstract

In recent decades, climate change has emerged as one of the most pressing global challenges, demanding urgent action. Türkiye, a country strategically positioned at the crossroads of Europe and Asia, has witnessed increasing concern over its vulnerability to the impacts of climate change and the need for transition toward a greener and sustainable future. This article examines Türkiye's noteworthy strides in addressing climate change both on the national and international stage. Beginning with an exploration of the context and background of climate change, the article delves into the country's climate vulnerabilities and the necessity for a comprehensive national agenda. Subsequently, it analyzes Türkiye's climate change policies and legislative developments, focusing on the states' climate change agenda, and legal and institutional frameworks. The key stakeholders and partnerships involved in climate action are also discussed. Furthermore, Türkiye's position in international climate diplomacy is scrutinized, including its participation in global climate summits, contributions to international climate agreements, and alignment with global climate goals. Ultimately, the article presents a comprehensive picture of Türkiye's political progress on climate change, highlighting both its achievements and challenges, and

[1]https://orcid.org/0009-0002-7327-7599
[2]https://orcid.org/0009-0009-9522-2063
[3]https://orcid.org/0000-0003-0312-065X

**The Social Consequences of Climate Change, 187–206**

**Published under exclusive licence by Emerald Publishing Limited**
**doi:10.1108/978-1-83797-677-520241014**

offering insights into the complexity of climate action and the potential for global collaboration in combatting the shared threat of climate change.

*Keywords*: Climate change; Türkiye; climate diplomacy; climate policies; climate action; cooperation

## Introduction

Climate change stands as one of the paramount challenges confronting humanity in the 21st century. Last decades have witnessed a steady rise in temperatures, the gradual melting of glaciers, the swelling of sea levels, and alterations in rainfall patterns across the globe. Consequently, the frequency of climate-related disasters has surged annually, resulting in numerous casualties and significant economic losses.

Although climate change is a natural phenomenon occurring over millennia, scientific researches indicate that human activities and population growth have expedited this process in recent times. According to the World Meteorological Organization (WMO), the five-year period from 2015 to 2019 is observed to be the warmest on record globally, with a 1.1°C increase in global temperatures compared to the preindustrial era and a 0.2°C rise compared to the previous five-year span (Bostanoğlu, 2020). This escalation underscores the inadequacy of current global climate initiatives in limiting the rise in global temperatures to below 2°C above preindustrial levels by 2,100, let alone the more stringent target of 1.5°C.

Addressing climate change demands collective action from all stakeholders, including individuals, governments, businesses, civil society organizations, and educational institutions. Key actors, particularly governments and private sectors, have already begun taking steps to tackle climate change and adapt to its adverse impacts. Given Türkiye's vulnerability to climate change, it is imperative for Türkiye to adopt ambitious climate policies. Addressing climate change through both mitigation and adaptation efforts remains a critical concern for Türkiye, a developing nation significantly exposed to climatic variation. Understanding Türkiye's historical and contemporary climate change policies and its position in international climate diplomacy is paramount, given its susceptibility to climate-related challenges.

## Understanding Türkiye's Climate Challenges

### *Climate Vulnerabilities and Risks in Türkiye*

Climate change poses a heightened threat to vulnerable systems and exacerbates pre-existing risks. Evaluating risks associated with climate and pinpointing corresponding significant vulnerabilities have become crucial aspects in effectively tackling the challenges posed by climate change. The third Assessment Report (AR3) of the Intergovernmental Panel on Climate Change (IPCC) defines vulnerability in the context of climate change as the degree to which a system is susceptible to, or unable to cope with, adverse effects of climate change, including climate variability and extremes. Vulnerability is a function of the character, magnitude, and rate of climate change and variation to which a system is exposed its sensitivity, and its adaptive capacity (IPCC, 2001, p. 398).

Türkiye, located within the Mediterranean Basin, is in one of the world's swiftest warming regions, rendering it exceptionally susceptible to climate-related hazards (MedECC, 2020). The repercussions of climate change on the Mediterranean Basin, including heightened risks of forest fires, droughts, heat waves, and altered precipitation patterns due to ocean acidification, are widely recognized (MedECC, 2020). This region has been identified as one of the globe's most vulnerable areas (Milano et al., 2013). Hence, the monitoring and forecasting of climate change phenomena has paramount importance in addressing the climate crisis, aiming to mitigate or, at the very least, alleviate its adverse impacts.

Recent studies indicate that temperatures have increased in almost all regions of Türkiye (Şen et al., 2017). The most increase is observed in the summer season while warm seasons expand. In the last 50 years, Türkiye's annual average temperatures have increased from 12.5°C to 14.5°C. According to the fourth Assessment Report of IPCC, the Mediterranean Basin will experience warmer temperatures, expanding drought, and increasing number of heat waves (IPCC, 2013). When regional projections are considered, the temperature in the eastern and southeastern parts of Türkiye is projected to increase by 6 °C by 2,100. Approximately 60% of the land area in Türkiye experiences an annual moisture deficit and is vulnerable to desertification (Türkeş, 2013). The forecasted rise in temperatures and alterations in precipitation patterns in certain regions of Türkiye are anticipated to heighten the frequency and severity of droughts. Consequently, Türkiye is expected to experience significantly drier climatic conditions in the future, rendering it exceptionally vulnerable to the impacts of climate change, particularly the escalation of drought occurrences (Kostopoulou & Jones, 2005).

Beyond shifts in average climate conditions, climate change also entails alterations in the frequency and severity of extreme weather and climate events. According to the IPCC, extreme weather or climate events are characterized by values of weather or climate variables surpassing or falling below threshold values near the upper or lower bounds of historical measurements. While individual climate events may not be extreme in themselves, their combined occurrence can lead to climate extremes such as droughts and floods. Climate change significantly influences alterations in the frequency, intensity, spatial extent, duration, and timing of these extremes (Seneviratne et al., 2012). There has been noticeable variability in climate extremes, including declines in cold days and nights, upticks in warm days and nights, longer durations of warm spells, and an increase in days with heavy precipitation (Jones et al., 2008).

Some extreme events are likely to become more frequent, longer lasting, more intense, and more widespread by 2,100 in Türkiye (Talu & Kocaman, 2019). Due to human impacts on climate change, the frequency of warm days in Türkiye is increasing and becoming more commonplace while the number of cold days is decreasing. A study focusing on changes in the annual count of record maximum and minimum air temperatures, using data from 81 stations across Türkiye spanning from 1950 to 2014, reveals a decline in the annual count of record minimum temperatures since the 1950s. Moreover, there's a notable upward trend in the frequency of record maximum temperatures, particularly evident after the 2000s. Notably, half of the record maximum events recorded since 1950 occurred during the period from 2000 to 2014 (Erlat & Türkeş, 2015). Therefore, it's imperative for Türkiye to urgently combat the impacts of climate change. The looming threat demands swift action through adaptation strategies and robust mitigation efforts to safeguard against its dire consequences.

## National Climate Policies and Legislative Developments

### *History of Legal and Institutional Framework for Environmental Policy*

Over the ensuing decades, Türkiye's environmental legislation has undergone iterative evolution, largely influenced by contemporary developments and changing perspectives. The enactment of Environment Law No. 2872 on August 9, 1983, marked a significant milestone as Türkiye's inaugural environmental legislation (Talu & Kocaman, 2019). Preceding this law, environmental protection measures were primarily administered by various ministries and institutions, each within their specific domains of interest. Following the implementation of the Environment Law, the General Directorate of Environment was established on June 8, 1984, to oversee its enforcement. Subsequently, a series of regulations and directives have been introduced and enforced over time.

Notably, the prospect of European Union (EU) membership has been instrumental in shaping Türkiye's environmental agenda. Negotiations for accession to the "Environment and Climate Change" chapter commenced on December 21, 2009. In the course of accession discussions, Türkiye has diligently worked to align its environmental regulations with the EU's environmental acquis. This alignment process has encompassed diverse areas such as air and water quality, environmental conservation, waste management, industrial emissions, chemical regulations, noise pollution, and, notably, climate change mitigation and adaptation efforts.

Initially, environmental policies within the development plans primarily aimed at addressing emerging pollution issues. However, they later progressed to include preventive measures and ultimately evolved into policies prioritizing the integration of the environment and economy in line with the concept of sustainable development (T.C. Başbakanlık Devlet Planlama Teşkilatı, 2000. *Uzun Vadeli Strateji ve 8. Beş Yıllık Kalkınma Planı*). Primary discussions regarding climate change has begun with the 8th Development Plan (2001–2005).

The 8th Development Plan established a dedicated Climate Change Specialization Commission during its preparation phase, although the plan itself only briefly addressed climate change. The preservation of the global climate system and Türkiye's responsibilities in combating climate change were tied to the principle of differentiated obligations, which remains valid. The plan aimed to increase energy efficiency and savings to control and reduce greenhouse gas emissions originating from the transportation, energy, industry, and housing sectors (T.C. Başbakanlık Devlet Planlama Teşkilatı, 2000. *Uzun Vadeli Strateji ve 8. Beş Yıllık Kalkınma Planı*). The 8th Development Plan outlined efforts toward becoming a party to the UNFCCC process alongside regulatory measures aimed at enhancing energy efficiency for greenhouse gas reduction (T.C. Çevre ve Şehircilik Bakanlığı, 2016).

The 9th Development Plan (2007–2013) allocated less space to climate change compared to the previous plan (T.C. Başbakanlık Devlet Planlama Teşkilatı, 2006. *Dokuzuncu Kalkınma Planı*). Prepared during a period when Türkiye's accession to the Kyoto Protocol was under discussion, the plan only mentioned fulfilling obligations arising from the UNFCCC under the title "Preservation of the Environment and Development of Urban Infrastructure". It indicated the preparation of a National Action Plan outlining policies and measures to reduce greenhouse gas emissions. In line with the 9th Development Plan, Türkiye has taken another step toward combating climate change by preparing a National Climate Action Plan outlining greenhouse gas reduction policies and measures tailored to its own conditions.

The 10th Development Plan (2014–2018) expanded the coverage of climate change topics under sections titled "Climate Change and Environment," "Effective Use of Food, Water, and Natural Resources," and "Livable Spaces, Sustainable Environment." The plan emphasized policies related to climate change and other environmental issues within the framework of sustainable development principles. It highlighted the development of environmental standards for competitiveness and green growth in production and consumption (T.C. Kalkınma Bakanlığı, 2014. *Onuncu Kalkınma Planı*).

The 11th Development Plan (2019–2023) frequently references climate change, indicating Türkiye's status as a developing country and its commitment to green growth and limiting the trend of emission increases. The plan underscores the importance of efforts to adapt to climate change, highlighting the potential for climate finance access to reduce marginal costs in emission reduction, particularly in industries (Türkiye Cumhuriyeti Cumhurbaşkanlığı Strateji ve Bütçe Başkanlığı, 2019. *On Birinci Kalkınma Planı*). It discusses the relationship between climate change and food security, as well as efficient water use, reflecting a resilient settlement approach to climate change.

The 12th Development Plan (2024–2028) was published in October 2023. Various aspects directly related to climate change addressed in the 12th Development Plan which includes (Türkiye Cumhuriyeti Cumhurbaşkanlığı Strateji ve Bütçe Başkanlığı, 2023. *On İkinci Kalkınma Planı*):

- The increasing prominence of low-carbon growth and green economy approaches, offering significant transformation opportunities for adaptation to climate change and emission reduction policies.
- The emergence of international climate justice policies, driven by the mechanisms of the Paris Agreement activating carbon pricing, green incentives, and regulations.
- Setting a net-zero emissions target for 2053 following the signing of the Paris Agreement in 2021 and updating the National Contribution Statement in 2022, along with the preparation of the Green Deal Action Plan in 2021 to comply with the European Green Deal.

- The prioritization of climate finance and support for emission reduction at the lowest cost for sectors affected by border carbon adjustment mechanisms, along with the implementation of the Emission Trading System.
- The potential for new opportunities in green jobs in Türkiye where green technologies are not sufficiently developed, accelerating green transformation in manufacturing industries, especially in emission-intensive sectors.
- The importance of combating climate change in energy supply security by promoting renewable energy, energy efficiency, energy storage, nuclear energy, and green hydrogen production, as well as accelerating energy transformation in buildings and increasing the use of geothermal heating.
- Considering the effects of climate change on food security and efficient water use, utilizing alternative water sources.
- Maintaining the importance of climate-friendly, sustainable, smart, calm, and resilient approaches such as resilient cities amid the increasing impact of climate change-related disasters, with the presence of the Sendai Framework for Disaster Risk Reduction.
- The impact of the 2053 net-zero emissions target on energy balance and the future of coal policies, emphasizing sustainable management of natural resources, energy transformation, green infrastructure, and circular economy investments without compromising development priorities.
- Promoting institutional capacity building to determine and mitigate the impact of climate change on health, as well as enhancing inter-agency collaboration.
- Increasing the adaptability of the tourism sector to climate change.
- Diversifying agricultural activities and ensuring effective and efficient use of agricultural lands to alleviate the effects of climate change and drought-induced poverty in rural areas.

To combat climate change and incorporate climate-related goals into its development agenda, Türkiye initially formulated the Climate Change Strategy. This strategy outlined a series of climate actions to be implemented over the short, medium, and long terms within the timeframe of 2010–2023. Subsequently, Türkiye introduced the Climate Change Action Plan and the National Climate Change Adaptation Strategy and Action Plan for the period spanning 2011–2023 (T.C. Çevre ve Şehircilik Bakanlığı, 2012. *Türkiye'nin İklim Değişikliği Uyum Stratejisi ve Eylem Planı*). The Climate Change Action Plan delineated various climate actions across multiple sectors, including buildings, transportation, industry, waste management, land use, forestry, and climate change adaptation measures. Concurrently, the National Climate Change Adaptation Strategy and Action Plan aimed to establish objectives pertaining to water resource management, the agricultural sector, food security, ecosystem preservation, biodiversity, forestry, disaster risk management, public health, and overarching issues related to adaptation efforts (T.C. Çevre ve Şehircilik Bakanlığı, 2012. *Türkiye'nin İklim Değişikliği Uyum Stratejisi ve Eylem Planı*). This plan outlines actions for controlling greenhouse gas emissions and adaptation efforts, defining responsible parties and timelines for implementing these actions, aligned with the objectives of the National Climate Change Strategy Document.

### *Development of Türkiye's Climate Change Agenda*

Governments worldwide are increasingly embracing policies and measures to tackle climate change, although progress varies significantly across nations. Each country experiences climate-related impacts at different magnitudes. Türkiye, with a population exceeding 84 million, stands out as one of the countries most severely affected by the adverse consequences of climate change (Bütün Bayındır, 2022). Given this context, the legal and institutional frameworks for climate change adaptation have become increasingly crucial for the country to address climate change and its ramifications.

The need for additional reforms at the global level has emerged to achieve economic growth while considering the climate agenda and encouraging investors and businesses to operate in alignment with scenarios that limit global warming. This necessity has brought the fight against climate change to the forefront of international economic and trade policies. The European Union, through its announcement of the European Green Deal on December 11, 2019, set the goal of becoming the first climate-neutral continent by 2050 (Fetting, 2020). Alongside, EU declared its adoption of a new growth strategy necessitating the transformation of its industries and reshaping all policies around climate change. In parallel with Türkiye's status as a developing nation, efforts toward emission reduction and climate adaptation continue. In this context, Türkiye has developed its own Green Deal Action Plan (T.C. Ticaret Bakanlığı, 2021. *Yeşil Mutabakat Eylem Planı*).

On October 29, 2021, a Presidential Decree led to the renaming of the Ministry of Environment and Urbanization to the Ministry of Environment, Urbanization, and Climate Change of Türkiye, becoming the highest level authority overseeing local and international efforts in mitigation, adaptation, and implementation (Presidential Decree No. 85, 2021). Simultaneously, the 'Directorate of Climate Change' and its associated 'Climate Change and Adaptation Coordination Board' were established by the same Decree. The Directorate is tasked with formulating plans, policies, and strategies related to climate change. It conducts studies to identify local, sectoral, and national adaptation needs, prepares modeling and risk assessment studies to gauge climate change impacts, and develops risk maps. It determines corresponding policies and facilitates cooperation and coordination between public and private sector entities and organizations.

Additionally, Türkiye, under the Ministry of Environment, Urbanization, and Climate Change, is establishing new institutions and carrying out social responsibility projects such as the Zero Waste Initiative, which aims to address the causes and potential solutions of climate change. The increasing focus on the impacts of climate change, particularly in the Mediterranean region, underscores the importance of efforts to prevent and mitigate its effects on human life and marine ecosystems.

Besides this, Türkiye's first Climate Council, a critical step in combating climate change, was held in Konya, one of the most climate-affected cities in Türkiye. Within this framework, 217 new decisions were made in line with the goal of achieving net zero emissions and green development by 2053 in the fight against climate change (T.C. Çevre, Şehircilik ve İklim Değişikliği Bakanlığı, 2022. *İklim Şurası'nda Alınan Önemli Kararlar*). The significant headings of the decisions included Climate-Adaptive Cities, Climate-Friendly Agriculture, Drought Action Plan, Environmental and Clean Transportation Network, Green Energy, Green Economy, and Climate Education.

Additionally, the intention the establish a Regional Activity Centre (RAC) focused on climate change, integrated into the United Nations Environment Program/Mediterranean Action Plan (UNEP/MAP) system under the Convention for the Protection of the Marine Environment and the Coastal Region of the Mediterranean (the Barcelona Convention) hosted by Türkiye was initially expressed during the 22nd Meeting of the Contracting Parties to the Barcelona Convention (COP 22) held in Antalya on December 7–10, 2021, to which Mrs Emine Erdoğan first lady of Türkiye has attended (Barcelona Convention Bulletin, 2021). Subsequently, Türkiye's proposal has introduced the elevation of the Mediterranean Climate Change Research Center (AIDAM) within the framework of Ankara University National Center for the Sea and Maritime Law (DEHUKAM) to the status of the Climate Change Regional Activity Centre (CC/RAC), integrated into the UNEP/MAP. The decision (UNEP/MAP IG. 26/12, 2023) regarding the establishment of the Climate Change Regional Activity Centre under the auspices of Türkiye was unanimously accepted at the 23rd Meeting of the Contracting Parties to the Barcelona Convention (COP23) held in Portoroz, Slovenia, from December 5th to 8th, 2023.

The establishment of the Climate Change Regional Activity Centre in Türkiye will not only strengthen Türkiye's position within the Barcelona Convention and the UNEP/MAP system but also enhance its active role on international platforms concerning climate change. In this context, the Centre aims to address the adverse effects of climate change on marine environments and to fulfill our responsibilities not only in marine but also terrestrial environments.

Given Türkiye's strategic geographical location surrounded by seas, possessing vital straits and rich inland water resources, it is imperative to utilize and preserve these areas to the maximum extent possible. Planning, developing strategies, and creating infrastructures are essential to achieve access to the power and position that can guide maritime policies in our region and address the impacts of climate change. The establishment of the CC/RAC will probably enhance Türkiye's effectiveness in the Mediterranean region and globally, as it holds significant geopolitical importance (Anadolu Agency, 2023. *Climate change center for Mediterranean to be established in Istanbul.*) Furthermore, it is noteworthy to emphasize Türkiye's dedication to addressing climate change and taking significant steps toward combating it. The establishment of the CC/RAC underscores Türkiye's commitment to actively participating in international efforts to mitigate the impacts of climate change.

### *The Need for a Comprehensive National Agenda*

In Turkish legislation, the regulation of climate change is not consolidated under a single law; instead, it is addressed horizontally across various legislative instruments. Direct references to climate change are found within the Environment Law, the Soil Protection and Land Use Law, and the Energy Efficiency Law. Additionally, greenhouse gas emissions are addressed within the Law on the Use of Renewable Energy Resources for Producing Electric Energy. The explicit mention of climate change is observed in the provisions of Presidential Decree Law No.1 and No.4, which govern the organization and authority of several ministries (Talu & Kocaman, 2019). However, the coverage of climate change in these laws and presidential decrees is not complex, detailed, or systematic; rather, it is contextual.

The overarching policies concerning climate change are largely influenced by international agreements such as the United Nations Framework Convention on Climate Change (UNFCCC), the Kyoto Protocol, and international agreements related to ozone layer protection and deforestation. There is ongoing debate among all stakeholders that are effecting from climate change regarding whether climate change should be regulated under a specific code of law.

The law would involve regulations to mitigate climate change, reduce its impacts, and facilitate adaptation if Turkish legislation were enacted to address climate change comprehensively. Such a law would cover the principles and procedures related to limiting activities contributing to climate change, mitigating greenhouse gas emissions, and adapting to climate change impacts. In this case, definitions of climate change-related concepts, identification of relevant public institutions and organizations, delineation of activities contributing to climate change, measures for mitigating greenhouse gas emissions, and protection and development of sink areas would be included. Enacting systematic legislation on climate change could elevate its status from being solely an environmental issue to a strategic consideration within public policies. It could also provide stability and enforceability to mitigation and adaptation policies by making them binding.

Türkiye has taken a significant stride in the realm of climate change by the Draft Climate Change Law (T.C. Çevre, Şehircilik ve İklim Değişikliği Bakanlığı, 2023. *İklim Değişikliği Kanunu Taslağı*). The primary objective of the Draft Climate Change Law is to realize Türkiye's vision of green development and achieve net zero emission targets. Within this framework, provisions have been outlined for reducing greenhouse gas emissions, undertaking climate adaptation activities, managing revenues, permits, and inspections, and establishing the necessary legal and institutional frameworks. The Draft Climate Change Law marks a strategic and pivotal step in the Türkiye's fight against climate change. Implementing the draft would enable Türkiye to move toward a sustainable future both domestically and internationally. This would hold paramount importance for the country's future, both economically and environmentally.

## International Climate Diplomacy and Türkiye

Addressing the challenges of climate change necessitates collaborative efforts on a global scale, underscoring the importance of collective action. Mitigating the impacts of climate change requires the active participation of all nations, irrespective of their developmental status. Key multilateral agreements such as the UNFCCC, the Kyoto Protocol, and the Paris Agreement serve as foundational frameworks in pursuit of these objectives. Türkiye plays a significant role in international climate diplomacy, actively engaging in global efforts to address climate change while advocating for its national interests and priorities. Türkiye's main policy in international climate diplomacy is based on climate justice and establish fairness among the Parties.

### *Türkiye's Contributions to International Climate Agenda*

Türkiye recognizes the importance of collective action in addressing climate change and has made significant contributions to international climate agreements and initiatives. With active participation in negotiations and making commitments to reduce greenhouse gas emissions, Türkiye demonstrates her willingness to be part of the global solution to climate change.

Türkiye has made pledges under international agreements such as the Paris Agreement, committing to reduce her greenhouse gas emissions and enhance climate resilience. Additionally, Türkiye participates in international initiatives aimed at promoting renewable energy, improving energy efficiency, promoting a zero-waste approach (United Nations Sustainable Development Goals Knowledge Platform, 2019. *Turkey's Sustainable Development Goals 2nd VNR*) and supporting sustainable development in line with global climate objectives.

#### *United Nations Framework Convention on Climate Change*

The landmark initiative in tackling the human-induced impact of global warming on climate change was the adoption of the UNFCCC during the 1992 United Nations Environment and Development Conference in Rio de Janeiro. Coming into effect on March 21, 1994, the UNFCCC counts over 190 signatories, including Türkiye.

At its core, the UNFCCC aims the stabilization of Green House Gas (GHG) concentrations in the atmosphere, thus preventing detrimental anthropogenic interference with the climate system. Parties to the Convention commit to reducing greenhouse gas emissions, fostering research and technology cooperation, and promoting the safeguarding of natural sinks. Central to the UNFCCC's framework is the principle of common but differentiated responsibilities and respective capabilities, which considers each country's developmental priorities, goals, and unique circumstances in emissions reduction efforts (Kuf, 2018).

This principle acknowledges that countries bear varying levels of responsibility for emissions, stemming from their development levels and historical contributions to greenhouse gas emissions since the industrial revolution. Accordingly, UNFCCC categorizes Parties into three groups (UNFCCC. *Parties & Observers*):

(1) Annex-I Parties: These nations are mandated to mitigate greenhouse gas emissions, safeguard and enhance sinks, and report on their climate action measures and emissions data. This category includes 42 countries and the EU.
(2) Annex-II Parties: In addition to the responsibilities of Annex-I Parties, these Parties are tasked with facilitating the transfer of environmentally friendly technologies to developing countries and providing financial support for technology access. There are 23 countries in this category, along with the EU.
(3) Non-annex Parties: While encouraged to collaborate on emissions reduction efforts, research, technology, and sink protection, these countries are not bound by the obligations imposed on Annex-I and Annex-II Parties. This category encompasses 154 countries.

Türkiye, initially listed in both Annex-I and Annex-II due to its Organization for Economic Co-operation and Development (OECD) membership (Organization for Economic Co-operation and Development, 2024. *Member Countries*), was later removed from Annex-II at the 7th Conference of Parties (COP) of the UNFCCC in 2001. Consequently, Türkiye remains solely in the Annex-I list (Republic of Türkiye Ministry of Foreign Affairs, 2024. *United Nations Framework Convention on Climate Change (UNFCCC) and the Kyoto Protocol*).

*The Kyoto Protocol*

The Kyoto Protocol, adopted on December 11, 1997, during the 3rd Conference of the Parties to the UNFCCC (COP3), emerged from negotiations initiated by contracting parties of the UNFCCC. Its aim is to obligate countries to lower their greenhouse gas emissions below specified levels. Its flexibility mechanisms based on market economy principles and the enforcement system for noncompliance distinguishes it from other international environmental agreements. Specifically, developed nations were mandated to achieve emission reductions by at least 55% compared to 1990 levels within a five-year period spanning from 2008 to 2012 (Yaşar & Yıldız, 2009). Despite setting these targets, the Protocol's effectiveness relied on nonbinding commitments to stabilize greenhouse gas emissions (Republic of Türkiye Ministry of Foreign Affairs, 2024. *United Nations Framework Convention on Climate Change (UNFCCC) and the Kyoto Protocol*).

As is well known, the Kyoto Protocol had not been ratified immediately by Türkiye due to its unique status within the UNFCCC framework. Initially included in both Annex I and Annex II of the UNFCCC alongside developed nations, Türkiye, as a member of the OECD (Organization for Economic Co-operation and Development, 2024. *Member Countries*), was required to commit emission reductions unconditionally under Annex I (United Nations Framework Convention on Climate Change. *Annex-I*), and provide financing and technology transfer to developing countries as per Annex II (United Nations Framework Convention on Climate Change. *Annex-II*) when the convention was adopted in 1992.

However, Türkiye identified herself as a developing country and contested the obligations imposed on developed Parties under the Convention (UNFCCC MISC/CP.3, 1997). Despite years of negotiations aimed at having her name removed from both Annex I and Annex II, Türkiye's requests were consistently denied. Seeking an alternative approach, Türkiye argued that she was in the early stages of industrialization and proposed to be removed solely from Annex II, invoking the principle of "common but differentiated responsibility" at the COP6 in 2000 (UNFCCC CP/5/Add.1, 2000).

This plea was eventually granted at COP7 in Marrakesh in 2001(UNFCCC 26/CP.7, 2001), resulting in Türkiye's removal from Annex II. Subsequently, Türkiye became a party to the UNFCCC as the 189th member on May 24, 2004, positioning itself differently from other Annex I Parties. Türkiye's formal accession to the Kyoto Protocol is on August 26, 2009 (UNFCCC, 2024. *Parties, Türkiye*). However, Türkiye did not ratify Annex B of the Protocol to avoid any mandatory emission reduction commitments (Republic of Türkiye Ministry of Environment, Urbanization and Climate Change Directorate of Climate Change, 2024. *Kyoto Protocol*).

### *The Paris Agreement*

The Paris Agreement (United Nations Framework Convention on Climate Change. Paris Agreement. 2015), adopted during COP21 of the UNFCCC in Paris on December 12, 2015, marks a significant milestone in global climate efforts as the first legally binding agreement of its kind. Coming into force on November 4, 2016, it sets the framework for climate action post-2020 with the aim of limiting the global average temperature increase to well below 2°C above preindustrial levels, with efforts to further limit it to 1.5°C (UNFCCC, 2024. *The Paris Agreement*). The nationally determined contributions (NDCs), which require all Parties to continuously enhance their efforts to mitigate emissions and regularly report on their progress, are central to the Agreement. A global stocktake every five years evaluates collective progress and informs future actions (Catala & Wyns, 2022).

Building upon the Kyoto Protocol, the Paris Agreement governs climate actions beyond 2020. Aligned with the UNFCCC principle of common but differentiated responsibilities and respective capabilities, the Agreement distinguishes Parties between developed, developing and least developed countries without referencing specific Annex lists (Arı, 2022) Developed countries are urged to lead with absolute emission reduction targets, while developing nations are encouraged to enhance their mitigation efforts and transition toward emission reduction or limitation targets, considering their unique circumstances.

Türkiye, underscoring her commitment to address climate change, signed the Paris Agreement on April 22, 2016, and it came into effect with the publication of the "Law on the Approval of the Paris Agreement" in the Official Gazette numbered 31,621, dated October 7, 2021 (Republic of Türkiye Ministry of Foreign Affairs, 2024. *Paris Agreement and Türkiye's Position*). Türkiye's accession to the Paris Agreement supports both its national and international policies aimed at addressing climate change. Signatory countries to the Paris Agreement are required to submit their NDCs to the UNFCCC Secretariat, outlining the emission reduction targets that Parties aim to achieve. Türkiye submitted her NDC before the agreement was signed in 2015, pledging to reduce greenhouse gas emissions by 21% by 2030 (UNFCCC, 2015. *The INDC of Turkey*).

### *Türkiye's Position on International Climate Policies*

Türkiye, as a developing country located in the Mediterranean region, is adversely affected by global climate crisis and contributed to tackle climate change according to its national capacity, reducing 1.4 billion tons of GHG emissions between 1990 and 2007 (T.C. Kalkınma Bakanlığı, 2014. *Onuncu Kalkınma Planı*). Türkiye began actively participating in climate change negotiations during the period of the Intergovernmental Negotiating Committee (INC), which was responsible for drafting the UNFCCC. The INC categorized countries for the UNFCCC based on institutional criteria. OECD members were designated to lead in climate change mitigation efforts, offering financial assistance to developing nations and facilitating technology transfer for climate-friendly initiatives. As a founding OECD member, Türkiye was listed directly in both Annex-I, comprising developed nations responsible for historical emissions, and Annex-II, comprising developed nations required to provide financial support for developing countries of the UNFCCC, tasked with emission reduction and technology and finance provision. However, Türkiye's developmental status differed from other OECD members, prompting its desire to be removed from the UNFCCC Annexes and reclassified as a non-Annex country (Arı, 2022).

Despite comprising only about 1% of global emissions, Türkiye faces challenges in negotiations due to its lower historical responsibility for emissions compared to certain OECD members. Although Türkiye's attempts to secure a more equitable position, her classification as an OECD member during the drafting and adoption of the UNFCCC in the early 1990s led to her listing among developed Parties in Annexes. However, Türkiye's requests for a fairer stance have encountered obstacles or rejections, reflecting the complexities inherent in multilateral climate agreements.

Türkiye's primary negotiation strategy on these platforms has been to elucidate her unique circumstances. Finally, in 2001, at COP7 in Marrakesh, some of Türkiye's requests were granted (UNFCCC 26/CP.7, 2001); she was removed from Annex II and designated as a party in Annex I with special circumstances. The COP7 decision urged other countries to acknowledge Türkiye's distinct position, setting her apart from other Annex-I Parties. These circumstances were further refined over time, with COP16 (UNFCCC 1/CP.16, 2010), COP17 (UNFCCC. 3/CP.17, 2011), and COP18 (UNFCCC. 1/CP.18, 2012) yielding three new decisions enabling Türkiye to access financial resources for mitigation, adaptation, technology development, and capacity building. The COP20 reaffirmed the final COP decisions regarding Türkiye's position in 2014. However, clear guidelines for Türkiye's utilization of the Green Climate Fund were not explicitly outlined in these COP decisions.

### *Alignment With Global Climate Goals*

Türkiye aligns her climate policies and actions with global climate goals, recognizing the importance of collective efforts to limit global temperature rise and mitigate the impacts of climate change. By integrating international best practices and standards into her domestic climate policies, Türkiye demonstrates her commitment to contributing to global efforts to achieve a sustainable and carbon-neutral future.

Türkiye's climate policies and strategies are aligned with the objectives of the Paris Agreement, including its national climate action plan and long-term decarbonization goals. By setting ambitious targets for renewable energy deployment, energy efficiency improvements, and emissions reduction, Türkiye illustrates her commitment to transition toward a more sustainable and resilient economy.

In this regard, Türkiye submitted her first Intended Nationally Determined Contribution (INDC) to the UNFCCC in 2015, aiming for up to a 21% reduction in greenhouse gas emissions by 2030 (UNFCCC, 2015. *The INDC of Turkey*). Türkiye updated its NDC at the 27th Conference of the Parties (COP) summit, on November 15, 2022 (UNFCCC, 2022. *Republic of Türkiye Updated First Nationally Determined Contribution*). As per the NDC declaration, Türkiye has revised her emission reduction target for 2030 from the initial 21% stated in the first INDC in 2015 to a heightened goal of 41%. Additionally, the statement asserts that Türkiye, classified as a developing nation, anticipates reaching its emissions peak by 2038 (T.C. Çevre, Şehircilik ve İklim Değişikliği Bakanlığı İklim Değişikliği Başkanlığı, 2022. *Türkiye Ulusal Katkı Beyanı'nı COP27'de Açıkladı*). The revised NDC encompasses a broad spectrum of sectors and activities, featuring comprehensive strategies for both mitigation and adaptation. It also provides insights into the mechanisms and resources required for implementation. Notably, the adaptation components are emphasized for the first time, focusing on sectors such as agriculture, forestry, water management, disaster risk reduction, urban and rural development, as well as public health (UNDP Climate Promise, 2023. Türkiye).

As a key regional player, Türkiye has the potential to exert influence and leadership on global climate issues. By demonstrating her commitment to climate action, promoting innovation and collaboration, and engaging constructively in international climate diplomacy, Türkiye can contribute shaping the global climate agenda and advancing the transition to a carbon-neutral future.

## Overcoming Challenges and Future Outlook

### *Evaluating the Effectiveness of Climate Policies of Türkiye*

Despite Türkiye increasingly feeling the impacts of the climate change, including rising temperatures, more frequent natural disasters, and drought, she has emerged as one of the countries with the fastest-growing GHG emissions due to her pursuit of national development objectives. Despite the presence of international instruments, Türkiye's emissions have continued to climb, underscoring the complexity of the climate change issue, with GHG emissions identified as the primary culprit. The insufficient emphasis on transitioning away from coal, a crucial aspect of combating climate change, could impede progress toward emission reduction targets. Countries must develop clear strategies and plans for transitioning to cleaner energy sources as part of comprehensive climate change policies.

Notably, Türkiye's NDCs fail to mention coal, while highlighting forests as critical GHG emission sink areas. However, specific details on widespread implementation of functional fire management plans, utilization of information technologies for forest protection, and enhanced measures for preventing forest fires are lacking.

The Paris Agreement marks a new phase in Türkiye's battle against the climate crisis. While responsibilities are shared, decision-makers hold the authority to take action. Urgent action is imperative to prevent escalating ecological, economic, and societal tragedies.

Despite the approval of the Paris Agreement in 2021, updated NDCs project an increase rather than a reduction in emissions, with greenhouse gas emissions expected to surge by over 30% by 2030 (Kovancı, 2023). Achieving the carbon-neutral goal by 2053 appears increasingly unlikely under these circumstances. Türkiye requires a specific climate change law to achieve its carbon neutrality objective, which entails recognizing the climate crisis as binding and establishing a regulatory framework to combat it. This framework shall include market-based mechanisms like emissions trading systems, science-based guidance, civil society involvement, transparent planning tools such as carbon budgets, financial and budgetary policies, and inter-agency coordination. Additionally, fundamental principles such as climate justice, human rights, transparency, accountability, social equity, nature conservation, and energy democracy must be acknowledged within this context. A climate change law is essential for ensuring climate justice in Türkiye.

Implementing mitigation and adaptation policies to combat climate change is crucial for Türkiye, especially considering her vulnerability to the impacts of global warming. The primary strategy involves reducing GHG emissions, notably those linked with fossil fuel consumption. Türkiye aims to bolster her energy security by increasing the use of renewable energy sources and domestic coal reserves for electricity generation.

Türkiye must align with the Paris Agreement by phasing out coal by 2030 and significantly increasing renewable energy generation. By mid-century, Türkiye should achieve complete decarbonization of her electricity sector to drive similar efforts in other sectors through electrification. In parallel, Türkiye's transportation sector needs electrification and emphasis on public transportation to reduce emissions by one-third until 2030, reaching net-zero emission levels by 2050, in line with Paris Agreement targets. Türkiye's aspiration to produce domestic electric vehicles aligns with global trends and enhances competitiveness.

## Conclusion

Climate change has emerged as a pressing global issue, demanding immediate attention and action from all nations. Certain regions, particularly those identified as vulnerable, bear the impacts of the climate crisis more acutely and must grapple with its repercussions more intensely. While the countries responsible for triggering the climate crisis may not always overlap with those most severely impacted by its effects, the response to climate change demands a comprehensive global approach.

In this context, the United Nations has taken a leading role in addressing climate change, initiating concrete measures to tackle this multifaceted challenge. The process, initiated under the UNFCCC, culminated in the establishment of three crucial pillars together with the Kyoto Protocol and the Paris Agreement. Regularly convened COP meetings serve as vital forums for assessing the progress of global efforts in combating climate change. These international accords and ongoing conferences play a crucial role in coordinating worldwide endeavors to mitigate the impacts of climate change.

This paper sheds light on the evolution of Türkiye's climate change policies over time, her level of engagement in international initiatives, and recent alignment with the goals outlined in the Paris Agreement. However, despite Türkiye's declaration of carbon neutrality goal for 2023, she remains evident that more concerted efforts are needed to attain this objective. Türkiye's continued reliance on fossil fuel-based policies and her limited progress in transitioning to renewable energy sources suggest a deviation from sustainability objectives. Therefore, a more comprehensive and accelerated approach is warranted to achieve Türkiye's carbon neutrality target and address the challenges posed by climate change effectively.

Nevertheless, Türkiye's active participation in international climate diplomacy, contributions to global climate instruments, and alignment with global climate goals underscore her commitment to address climate change as a responsible member of the international community.

## References

Anadolu Agency. (2023). Climate change center for Mediterranean to be established in Istanbul. https://www.aa.com.tr/en/environment/climate-change-center-for-mediterranean-to-be-established-in-istanbul/3086541#:~:text=The%20Climate%20Change%20Regional%20Activity,ensuring%20regional%20adaptation%20to%20climate

Arı, İ. (2022, Spring). Türkiye's climate policy. *Insight Turkey*, *24*(No. 2), 89–112. https://doi.org/10.25253/99.2022242.6

Barcelona Convention Bulletin, International Institute for Sustainable Development. (2021). Summary of the 22nd meeting of the contracting parties to the Barcelona convention its protocols: 7-10 December 2021. https://enb.iisd.org/sites/default/files/2021-12/barcelona_convention_cop22.pdf

Bostanoğlu, M. N. (2020). *Where does Turkey stand on the international climate regime?* Economic Development Foundation. (IKV) Brief. https://www.ikv.org.tr/images/files/Where_Does_Turkey_Stand_on_the_International_Climate_Regime.pdf

Bütün Bayındır, G. D. (2022). *Development of a climate risk assessment method for the provinces of Türkiye*. The Graduate School of Natural and Applied Sciences of Middle East Technical University. https://open.metu.edu.tr/handle/11511/99404

Catala, A. P., & Wyns, A. (2022). The global stocktake: A health check for the Paris Agreement. *The Lancet Planetary Health*, *6*(4), 297–298. https://doi.org/10.1016/S2542-5196(22)00066-3

Erlat, E., & Türkeş, M. (2015). Türkiye rekor maksimum ve minimum hava sıcaklıklarının frekanslarinda 1950-2014 döneminde gözlenen değişmeler ve atmosfer koşullarıyla bağlantıları. *Ege Coğrafya Dergisi*, *24*(2), 29–55.

Fetting, C. (2020). The European Green Deal. *ESDN Report*. https://www.esdn.eu/fileadmin/ESDN_Reports/ESDN_Report_2_2020.pdf

IPCC. (2001). Climate change 2001: Synthesis report. *A contribution of working groups I, II, and III to the third assessment report of the Intergovernmental Panel on Climate Change* (p. 398). Cambridge University Press.

IPCC. (2013). Climate change 2013: The physical science basis. *A contribution of working group I to the fifth assessment report of the Intergovernmental Panel on Climate Change*. Cambridge University Press.

Jones, G. S., Stott, P. A., & Christidis, N. (2008). Human contribution to rapidly increasing frequency of very warm Northern Hemisphere summers. *Journal of Geophysical Research*, *113*(D2). https://doi.org/10.1029/2007jd008914

Kalkınma Bakanlığı, T. C. (2014). 10. Kalkınma Planı. https://www.sbb.gov.tr/wp-content/uploads/2022/08/Onuncu_Kalkinma_Plani-2014-2018.pdf

Kostopoulou, E., & Jones, P. D. (2005). Assessment of climate extremes in the Eastern Mediterranean. *Meteorology and Atmospheric Physics*, *89*(1–4), 69–85. https://doi.org/10.1007/s00703-005-0122-2

Kovancı, E. (2023). International climate change policies of Türkiye in its 100th year. *Akademik Hassasiyetler*, *10*(Cumhuriyet Özel Sayısı), 319–345. https://doi.org/10.58884/akademik-hassasiyetler.1350652

Kuf, K. H. (2018). The law of climate change mitigation: An overview. *Encyclopedia of the Anthropocene*, *2*, 505–510. https://doi.org/10.1016/B978-0-12-809665-9.10027-8

MedECC. (2020). Climate and environmental change in the Mediterranean Basin-current situation and risks for the future. *First Mediterranean Assessment Report*. (Version 1).

Milano, M., Ruelland, D., Fernandez, S., Dezetter, A., Fabre, J., Servat, E., Fritsch, J. M., Ardoin-Bardin, S., & Thivet, G. (2013). Current state of Mediterranean water resources and future trends under climatic and anthropogenic changes.

*Hydrological Sciences Journal*, *58*(3), 498–518. https://doi.org/10.1080/02626667.2013.774458

Organization for Economic Co-operation and Development. (2024). Member countries. https://www.oecd.org/about/members-and-partners/

Presidential Decree No. 85. (2021). https://www.resmigazete.gov.tr/eskiler/2021/10/20211029.pdf

Republic of Türkiye Ministry of Environment, Urbanization and Climate Change Directorate of Climate Change. (2024). *Kyoto Protocol*. https://iklim.gov.tr/en/kyoto-protocol-i-118

Republic of Türkiye Ministry of Foreign Affairs. (2024). *Paris Agreement and Türkiye's Position*. (2024). https://www.mfa.gov.tr/united-nations-framework-convention-on-climate-change-_unfccc_and-the-kyoto-protocol.en.mfa

Republic of Türkiye Ministry of Foreign Affairs. (2024). *United Nations Framework Convention on Climate Change (UNFCCC) and the Kyoto Protocol*. https://www.mfa.gov.tr/united-nations-framework-convention-on-climate-change-_unfccc_-and-the-kyoto-protocol.en.mfa

Şen, O., Bozkurt, D., Göktürk, O. M., Dündar, B., & Altürk. (2017). Türkiye'de iklim değişikliği ve olası etkileri. 3. Ulusal Taşkın Sempozyumu, İstanbul, Türkiye.

Seneviratne, S. I., Nicholls, N., Easterling, D., Goodess, C. M., Kanae, S., Kossin, J., Luo, Y., Marengo, J., McInnes, K., Rahimi, M., Reichstein, M., Sorteberg, A., Vera, C., & Zhang, X. (2012). Changes in climate extremes and their impacts on the natural physical environment. In C. B. Field, V. Barros, T. F. Stocker, D. Qin, D. J. Dokken, K. L. Ebi, M. D. Mastrandrea, K. J. Mach, G.-K. Plattner, S. K. Allen, M. Tignor, & P. M. Midgley (Eds.), *Managing the risks of extreme events and disasters to advance climate change adaptation. A special Report of working groups I and II of the intergovernmental panel on climate change* (pp. 109–230). Cambridge University Press.

Talu, N., & Kocaman, H. (2019). Turkey's climate change policy, legal and institutional framework. https://www.iklimin.org/wp-content/uploads/2020/03/modul_04_en-1.pdf

T.C. Başbakanlık Devlet Planlama Teşkilatı. (2006). Dokuzuncu Kalkınma Planı. https://www.sbb.gov.tr/wp-content/uploads/2022/07/Dokuzuncu_Kalkinma_Plani-2007-2013.pdf

T.C. Çevre ve Şehircilik Bakanlığı. (2012). İklim Değişikliği Uyum Stratejisi ve Eylem Planı 2011-2023. https://webdosya.csb.gov.tr/db/iklim/editordosya/uyum_stratejisi_eylem_plani_TR.pdf

T.C. Çevre ve Şehircilik Bakanlığı. (2012). İklim Değişikliği Ulusal Eylem Planı 2011-2023. https://webdosya.csb.gov.tr/db/iklim/banner/banner591.pdf

T.C. Çevre ve Şehircilik Bakanlığı. (2016). *Türkiye İklim Değişikliği 6. Ulusal Bildirimi*. https://webdosya.csb.gov.tr/db/destek/editordosya/Turkiye_Iklim_Degisikligi_Altinci_Ulusal_Bildirimi.pdf

T. C. Çevre, Şehircilik ve İklim Değişikliği Bakanlığı İklim Değişikliği Başkanlığı. (2022). *Türkiye Ulusal Katkı Beyanı'nı COP27'de Açıkladı*. https://iklim.gov.tr/turkiye-ulusal-katki-beyani-ni-cop27-de-acikladi-haber-84

T. C. Çevre, Şehircilik ve İklim Değişikliği Bakanlığı. (2022). *İklim Şurası'nda Alınan Önemli Kararlar*. https://www.csb.gov.tr/iklim-surasi-nda-alinan-onemli-kararlar-bakanlik-faaliyetleri-34154

Türkeş, M. (2013). İklim verileri kullanılarak Türkiye'nin çölleşme haritası dokümanı hazırlanması raporu. *Orman ve Su İşleri Bakanlığı, Çölleşme ve Erozyonla Mücadele Genel Müdürlüğü Yayını.* https://www.researchgate.net/publication/293334692_Iklim_Verileri_Kullanilarak_Turkiye%27nin_Collesme_Haritasi_Dokumani_Hazirlanmasi_Raporu

Türkiye Cumhuriyeti Başbakanlık Devlet Planlama Teşkilatı. (2000). Uzun Vadeli Strateji ve 8. Beş Yıllık Kalkınma Planı. http://www.bilgitoplumu.gov.tr/wp-content/uploads/2015/01/Sekizinci_Kalkinma_Plani.pdf

Türkiye Cumhuriyeti Cumhurbaşkanlığı Strateji ve Bütçe Başkanlığı. (2019). On Birinci Kalkınma Planı. https://www.sbb.gov.tr/wp-content/uploads/2022/07/On_Birinci_Kalkinma_Plani-2019-2023.pdf

Türkiye Cumhuriyeti Cumhurbaşkanlığı Strateji ve Bütçe Başkanlığı. (2023). On İkinci Kalkınma Planı. https://www.sbb.gov.tr/wp-content/uploads/2023/12/On-Ikinci-Kalkinma-Plani_2024-2028_11122023.pdf

Türkiye Cumhuriyeti Ticaret Bakanlığı. (2021). Yeşil Mutabakat Eylem Planı. https://ticaret.gov.tr/data/60f1200013b876eb28421b23/MUTABAKAT%20YE%C5%9E%C4%B0L.pdf

UNDP Climate Promise. (2023). Türkiye. https://climatepromise.undp.org/what-we-do/where-we-work/turkiye

UNEP/MAP. IG.26/12. (2023). 23rd meeting of the contracting parties to the Barcelona convention (COP 23). https://enb.iisd.org/sites/default/files/2023-12/barcelona_convention_cop23.pdf

UNFCCC. (2024). The Paris Agreement. https://unfccc.int/process-and-meetings/the-paris-agreement

UNFCCC. (2015). The INDC of Türkiye. https://unfccc.int/sites/default/files/NDC/2022-06/The_INDC_of_TURKEY_v.15.19.30.pdf

UNFCCC. (2022). Republic of Türkiye updated first nationally determined contribution. https://unfccc.int/sites/default/files/NDC/202304/T%C3%9CRK%C4%B0YE_UPDATED%201st%20NDC_EN.pdf

UNFCCC. 1/CP.16. (2010). 16th meeting of the contracting parties to UNFCCC (COP16). https://unfccc.int/resource/docs/2010/cop16/eng/07a01.pdf

UNFCCC. 1/CP.18. (2012). 18th meeting of the contracting parties to the UNFCCC (COP18). https://unfccc.int/resource/docs/2012/cop18/eng/08a01.pdf#page=3

UNFCCC. 26/CP.7. (2002). 7th meeting of the contracting parties to the UNFCCC (COP7). https://unfccc.int/resource/docs/cop7/13a04.pdf

UNFCCC. 3/CP.17. (2011). 17th meeting of the contracting parties to UNFCCC (COP17). http://www.ciesin.columbia.edu/repository/entri/docs/cop/FCCC_COP17_dec03.pdf

UNFCCC. CP/1997/Misc. (1997). Submission by Turkey. https://unfccc.int/cop5/resource/docs/cop3/misc03.htm

UNFCCC. CP/2000/5/Add.1. (2001). *Report of the conference of the parties on the first part of its sixth session, held at the hague from 13 to 25 November 2000. 23*. https://unfccc.int/resource/docs/cop6/05a01.pdf

UNFCCC. (2024). Parties. https://unfccc.int/node/61221

United Nations Framework Convention on Climate Change. (1992). https://unfccc.int/resource/docs/convkp/conveng.pdf
United Nations Sustainable Development Goals Knowledge Platform. (2019). Turkey's sustainable development goals 2nd VNR. https://sustainabledevelopment.un.org/content/documents/23862Turkey_VNR_110719.pdf
Yaşar, D., & Yıldız, D. (2009). *Küresel ısıtılan dünya ve su*. Truva Yayınları.

Printed and bound by CPI Group (UK) Ltd, Croydon, CR0 4YY

30/10/2024

14583693-0002